News for a Mobile-First Consumer

This book is part of the Peter Lang Media and Communication list.
Every volume is peer reviewed and meets
the highest quality standards for content and production.

PETER LANG
New York • Bern • Frankfurt • Berlin
Brussels • Vienna • Oxford • Warsaw

Paula M. Poindexter

News for a Mobile-First Consumer

PETER LANG
New York • Bern • Frankfurt • Berlin
Brussels • Vienna • Oxford • Warsaw

Library of Congress Cataloging-in-Publication Data

Names: Poindexter, Paula Maurie, author.
Title: News for a mobile-first consumer / Paula M. Poindexter.
Description: New York: Peter Lang, 2016.
Includes bibliographical references and index.
Identifiers: LCCN 2016014959 | ISBN 978-1-4331-2841-7 (hardcover: alk. paper)
ISBN 978-1-4331-2840-0 (paperback: alk. paper) | ISBN 978-1-4539-1815-9 (e-book)
ISBN 978-1-4331-3561-3 (epub) | ISBN 978-1-4331-3562-0 (mobi)
Subjects: LCSH: Journalism—United States—History—21st century.
News audiences—United States—History—21st century.
Journalism—Technological innovations—United States.
Electronic news gathering—United States—History.
Social media—United States—History. | Generation Y.
Classification: LCC PN4867.2 .P65 2016 | DDC 071/.3—dc23
LC record available at https://lccn.loc.gov/2016014959

Bibliographic information published by **Die Deutsche Nationalbibliothek**.
Die Deutsche Nationalbibliothek lists this publication in the "Deutsche Nationalbibliografie"; detailed bibliographic data are available on the Internet at http://dnb.d-nb.de/.

29 Broadway, 18th floor, New York, NY 10006
www.peterlang.com

To my daughter, Alexandra Wilson

TABLE OF CONTENTS

LIST OF FIGURES

LIST OF TABLES

LIST OF APPENDICES

PREFACE

More than half a century ago, when news was only available at appointed times from newspapers, TV, radio, and news magazines, almost 9 out of 10 adults in the U.S. read a newspaper every day. Today, with every type of news available on a mobile device anytime and from wherever consumers may be, fewer than half of adults in the U.S. seek news daily. Why are fewer people seeking news at a time when an unlimited amount of news is available on smartphones and tablets? And what, if anything, can be done to increase news consumption?

Increasing news consumption will not be an easy task because a new type of consumer has emerged in the news media landscape, replacing yesterday's newspaper and news magazine readers, TV and cable news viewers, radio news listeners, and online news consumers. Unfortunately, a fraction of what we know about pre-mobile news consumers is relevant to today's mobile consumers, leaving us with little insight about how to get more mobile consumers to engage with news on their smartphones and tablets.

Recognizing the need for new knowledge about the mobile consumer that has emerged in the news media landscape, I conducted a national survey to answer questions about mobile device owners, what they use their mobile devices for, and their news interests, engagement, and attitudes. Additionally,

I asked survey participants about mobile news, news apps, and their perceptions of journalists and the quality of U.S. journalism. The results, which are reported in *News for a Mobile-First Consumer*, call into question much of what we know about news consumers and news engagement when examined from a mobile perspective.

In addition to the discovery of three distinct mobile consumer types that populate the mobile landscape, the results suggest news organizations do not have unlimited time to figure out how to make news important to the lives and screens of mobile consumers. The survey results also underscore that journalism's future is dependent upon demographic groups that it has alienated because of the way they have been covered in the past. These demographic groups, which include Millennials, African Americans, Latinos, Asian Americans, Native Americans, and women, are at the forefront of the mobile revolution, so increasing news consumption must begin with them. Political leanings may not be as influential as one's generation, race, ethnicity, or gender in mobile news consumption, but they do matter in the mobile landscape, especially during a presidential election year. That's why *News for a Mobile-First Consumer* also includes data about political party, ideology, and mobile news engagement.

The national survey also provides new insight into social media as an integral part of the mobile landscape that can be both an ally and threat to news. A supplemental national survey digs deeper into social media by asking consumers about the news they have seen on social networking sites. With so much fanfare about Snapchat's potential as a platform to connect the youngest Millennials with news, Snapchat users are also asked about news.

While most of *News for a Mobile-First Consumer* examines mobile from the perspective of consumers, the book does not neglect journalism, including the news that news organizations deliver on smartphones, and why some of it may be the reason some consumers are not that enamored with mobile news. The survey data also re-affirm that journalism has a problem with trust; the public remains skeptical that journalism consistently serves the public good.

Just as Kovach and Rosenstiel proposed *The Elements of Journalism* to address what was seriously wrong with journalism, I have proposed the "Essentials of Mobile Journalism," a plan for journalism and news to thrive in the mobile age. These mobile journalism essentials, which are based on a synthesis of the results of the News for a Mobile-First Consumer Survey; the supplemental survey; national reports; reading news exclusively on my smartphone; as well as discussions with journalists, media professionals, and Millennial mobile users, will help news organizations produce news that will resonate in consumers' mobile lives.

Throughout the book I use my knowledge of journalism history and theories; my news media, business, research, teaching, and leadership experience; as well as my experience as a mobile news consumer to provide insight into today's mobile landscape and the consumers who inhabit it. I also call on my past research on Millennials, racial and ethnic groups, women, and social media to help me interpret some of the data reported in *News for a Mobile-First Consumer*.

Finally, as someone who has researched news consumers and non-news consumers and developed and implemented initiatives to encourage news consumption in a pre-mobile world and today's mobile era, I am confident that *News for a Mobile-First Consumer* will be indispensable to media scholars; journalists and media executives; thought leaders; and students who have yet to fully understand mobile consumers, their impact on the future of mobile news, and how journalism can fulfill its responsibility in a mobile and social media world that is nothing like the past.

ACKNOWLEDGMENTS

First and foremost, I would like to thank Mary Savigar, Peter Lang Publishing senior acquisitions editor, for her enthusiastic "yes" to *News for a Mobile-First Consumer*. When I proposed the book project to her, there was virtually no research literature about mobile news consumers and there was only one academic study about journalists who were using mobile devices to report the news. Special appreciation is also expressed to Peter Lang's Phyllis Korper, Bernadette Shade, Sophie Appel, who designed the eye-catching cover, and reviewers, and the copy editor who helped improve *News for a Mobile-First Consumer*.

Additionally, I want to acknowledge ONA, the Online News Association, because their inclusion of a mobile news panel at their 2013 annual conference which I attended, helped me make the case for a book on this topic and for my course proposal to then–University of Texas at Austin School of Journalism director Glenn Frankel. My proposed course News for a Mobile Audience, which I first taught spring 2014, was the first mobile news course taught exclusively to University of Texas at Austin journalism students. The other course on mobile apps, which is taught by my colleague Robert Quigley, is cross-listed with computer science.

My News for a Mobile Audience course was critical to the development of this book because it underscored the need for a book that focused on the transformed mobile landscape and the role of mobile news consumers, who are different from traditional news consumers. For that I thank the first students who enrolled in the class, spring semester 2014. Special recognition is also given to my 2016 spring semester students who used draft chapters of *News for a Mobile-First Consumer* and provided thoughtful feedback and insight.

I also want to acknowledge and thank the following people who have spoken to my class in person or via Skype: Eric Bishop, the mobile homepage editor of the *New York Times*; R. B. Brenner, former *Washington Post* Sunday editor, deputy universal news editor, and metro editor, who is now director of the University of Texas School of Journalism; Nicki Purcell, chief digital officer and senior vice president of consumer sales, *Dallas Morning* News; Tom Kent, deputy managing editor/standards editor, the Associated Press and chair of ONA's Build Your Own Ethics Code Project; Logan Molyneux, now a Temple University journalism professor; and Amy Zerba, senior staff editor at the *New York Times*, who has mastered the art of incorporating social media in journalism practice. By sharing their knowledge about journalism and consumers in the mobile landscape with students in my class, they have also provided me with insight that has enhanced *News for a Mobile-First Consumer*.

A special acknowledgment is also given to former *New York Times* executive editor Jill Abramson whom I invited to give the keynote address at the 2014 AEJMC conference in Montreal when I was president of this association of almost 4,000 journalism and communication educators, journalism and media professionals, and graduate students. Both in her official keynote speech and our informal conversation prior to the speech I had an opportunity to learn more about the mobile app NYT Now, which she was instrumental in helping the *New York Times* create.

Veronica Inchauste and O'Neil Provost of the then Office of Survey Research at the University of Texas at Austin also deserve acknowledgment because I commissioned them to conduct the News for a Mobile-First Consumer Survey which is the foundation for this book.

I also want to acknowledge AEJMC executive director Jennifer McGill and the AEJMC Board for supporting my presidential initiatives that were created to have a positive impact on increasing news engagement, understanding more about the news audience, and increasing awareness of journalistic excellence in this mobile-transformed news landscape. These initiatives include News Engagement Day, the News Audience Research Paper Award, and the Center for News Excellence and Engagement.

I am indebted to my Millennial daughter Alexandra Wilson to whom *News for a Mobile-First Consumer* is dedicated. As a mobile-first consumer, Millennial, and enthusiast of both social media and news, Alexandra has been an invaluable resource and patient listener as I shared with her the process and pressures of writing *News for a Mobile-First Consumer*.

Finally, as a teacher of journalism, student of journalism, and critic of journalism, I would like to thank my former professors, former and current students, and former and current colleagues, as well as the known and less well-known journalists and thought leaders who have contributed to, inspired, and shaped not just my teaching, research, writings, and critiques on journalism but also *News for a Mobile-First Consumer*, which is not just about news and consumers in the transformed mobile landscape but also about how journalism can fulfill its promise in the mobile world in which we now live.

· 1 ·

THE SMART MOBILE LANDSCAPE

Two years after Steve Jobs unveiled the iPhone, a Pulitzer Prize–winning editor from the *Washington Post* showed off his iPhone to a journalism class at the University of Texas at Austin. The editor, who was a journalist in residence that spring 2009 semester, was one of 12 invited speakers in the Journalism, Society, and the Citizen Journalist class that had been created as part of the Carnegie-Knight initiative on the future of journalism education. The class theme for that semester was "Conversations about Challenges Facing the Future of Journalism."

While the professor who created the class expected invited speakers to discuss challenges related to the collapse of the newspaper business model, newsroom layoffs, multimedia skills required of journalists, new competition (bloggers, citizen journalists, online-only news sites) in the media landscape, and the declining news audience, she had not expected that a new type of mobile phone would be part of the conversation. It was, perhaps, hard for the professor to fully grasp the iPhone's role in journalism's future because she—like most people in 2009—did not own one. But *The Post* editor did, which is why he showed off his iPhone with a screen filled with apps to the classroom of future journalists.

When the iPhone was released, it was unlike any mobile phone available to consumers. And even though the word "smartphone" had been coined almost two decades before (McCarty, 2011), it was the iPhone's features and operating system that made the mobile phone not just smart but also easy to use and fun. And the phone had a coolness factor, making it irresistible.

The iPhone and similar phones based on Google's Android operating system are no doubt the reason consumers have adopted smartphones at a faster rate than most other technologies, including landline phones, radio, computers, and the Internet. Landline phones, for example, took "about 45 years to get from 5 percent to 50 percent among U.S. households" while smartphones took four years to go from 5% to 40% penetration. Television was the only other technology in the U.S. to be adopted as rapidly as smartphones and that was 60 years ago (DeGusta, 2012).

By January 2014, mobile device ownership in the U.S., including smartphones, tablet computers, cellphones, and e-readers, had reached near saturation, according to a survey conducted by the Pew Research Center (2014). That year, nine out of ten adults owned cellphones; tablet computers were owned by two-fifths (42%) of U.S. adults and one-third (32%) owned e-readers. Ten months later, 64% of Pew Research Center survey participants owned a smartphone, which was the same percentage of smartphone owners in the national survey conducted for *News for a Mobile-First Consumer*. And over half (51%) of U.S. adults now owned tablets, according to the survey conducted for this book.

Further evidence of the saturation of mobile device ownership in the U.S. can be found by comparing landline phones and mobile phone ownership. According to the News for a Mobile-First Consumer Survey, 49% of survey participants owned both a landline and mobile phone and 44% owned a smartphone or cellphone only. Once dominant landline phones had declined so much that only 6% of survey participants said they only owned a landline phone.

Brief Description of Book's Research Method

Results from the News for a Mobile-First Consumer Survey are the foundation for this book. As explained in Appendix A, the national sample for the online survey was acquired from Survey Sampling, Inc., an internationally respected survey sampling firm. The survey questionnaire, which took 10 to 15 minutes to answer, was completed by 1,506 respondents. As the questionnaire in the appendix shows, the survey questions were written to capture mobile use patterns, news consumption, mobile news and social media use, attitudes toward journalism, and political and civic activity. Special attention in the design of the questionnaire and the data collection was paid to ensuring that the demographics of the participants in the online survey matched U.S. racial, ethnic, and gender demographics. Results showed the online survey demographics matched well with the U.S. Census: 62% White (Non-Hispanic), 19% Hispanic/Latino, 13% African American/Black, 5% Asian American, 1% Native American, and 52% female and 48% male. Frequencies, crosstabulations, factor analysis, and index construction were used to analyze the data and appropriate statistical tests were applied. The vast majority of correlations that are reported in the tables and text are statistically significant; occasionally non-significant correlations are reported because of the insight that can accompany non-significant results. In those cases, the results are labeled accordingly.

In addition to results from the News for a Mobile-First Consumer Survey, data from other surveys and reports will be referenced from time to time, but they will be clearly sourced. Finally, a supplemental survey was conducted June 2015 to gather information about news and Snapchat, a social media app that emerged in the mobile landscape after the main survey was conducted. The supplemental survey, which is labeled when its results are reported, also asks questions to determine how engaging with news on mobile devices differs from engaging with news on traditional platforms. The questionnaire and methodology for the supplemental survey can also be found in the appendix.

Smartphone Penetration beyond the U.S.

As fast as the adoption rate of smartphones has been in the U.S., the U.S. is not even among the top 10 countries in smartphone penetration. In 2014, the U.S. ranked 18th in world smartphone penetration behind Singapore, South Korea, Sweden, Hong Kong, Spain, China, Denmark, the UK, Norway, Taiwan, Australia, the Netherlands, Ireland, Israel, Switzerland, New Zealand, and Finland (Global Smartphone Penetration, 2014). China ranks sixth in smartphone penetration, with 557 million Chinese using smartphones and tablets to connect to the Internet, yet China has more mobile device owners than any country in the world (McKirdy, 2015). However, in Africa, where mobile phones are cheap, Africa is expected to have 1 billion mobile phones in 2016, which would surpass China (Veselinovic & Clements, 2015).

The mobile transformation of the world we live in appears to have happened overnight but the reality is this mobile landscape is the bi-product of the evolution of multiple innovations—computer, Internet, World Wide Web, microchip, search engines, digital technology, high-speed cellular, wireless, Android, apps, etc.—converging into a single sleek device that fits in a pocket or purse and effortlessly connects people, content, services, and commerce anytime and from wherever consumers may happen to be. As the Boston Consulting Group said in a 2015 report: "Mobile technologies have transformed the way we live, work, learn, travel, shop, and stay connected" (The Mobile Revolution, 2015). Yet, all consumers in this new mobile world are not the same.

Primary Types of Mobile Consumers

When mobile device owners were asked in the News for a Mobile-First Consumer Survey what they most used their mobile device for, their answers revealed three primary types of mobile consumers (see Figure 1.1.). Consumers who said they used a mobile device for *everything* were labeled "mobile-first" while consumers who said they mostly used their mobile devices for communication were labeled "mobile laggards." "Mobile specialist" was the third type of mobile consumer because the mobile device was primarily used for a specific purpose, whether connecting to certain types of content or engaging in specific activities.

These three types of consumers are not just different in their primary mobile device use, they also are different in their presence in the mobile landscape; at 66%, mobile laggards are the largest, and mobile-first, with only 2%, are the smallest. Almost one-third (32%) of mobile device owners are mobile specialists. This mobile consumer distribution is a snapshot of the current mobile landscape that is expected to change over the next decade as mobile-first consumers and mobile specialists increase and mobile laggards decline. Much of the increase in mobile-first and mobile specialists will be due to a generational shift as the oldest generations age and die and the post-Millennial generation grows up.

Who are these mobile consumer types? What socio-economic and political attributes, if any, distinguish them? An analysis of seven socio-economic and political variables found more similarities than differences among both mobile-first consumers and mobile laggards. In fact, an analysis by generation, race and ethnicity, gender, income, education, political party, and ideology

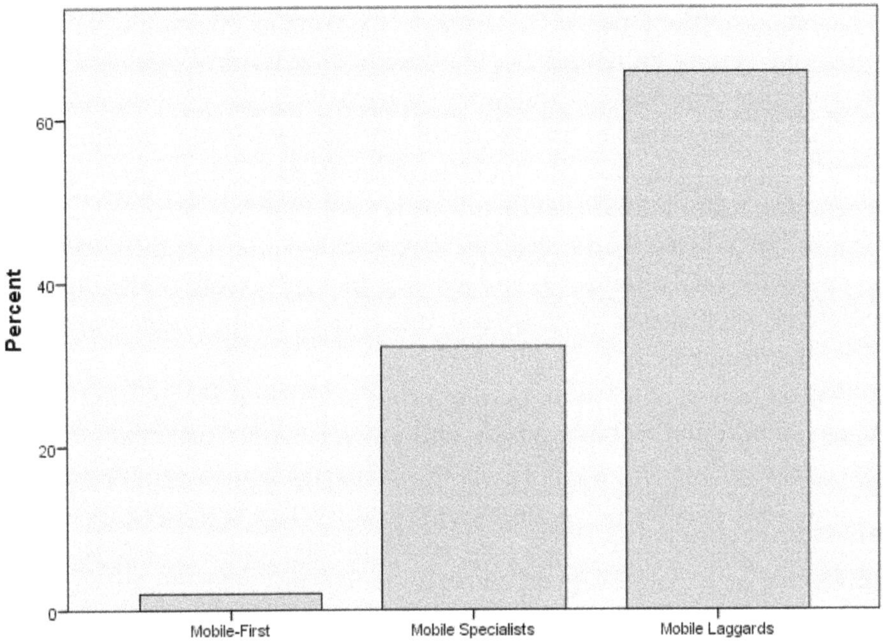

Figure 1.1: Mobile Consumer Types.

determined that only two variables—generation and race and ethnicity—distinguished consumers categorized as mobile-first and mobile laggards.

Millennials, the generation born between the early 1980s and late 1990s, are leading the way in being mobile-first consumers, while Baby Boomers, born between 1946 and 1964, are more likely to be mobile laggards. Sandwiched between Millennials and Baby Boomers is Generation X, born between 1965 and approximately 1982. Generation X has yet to have the impact of the older Baby Boomer Generation but they are influential as parents of younger Millennials who will continue the mobile transformation of the media landscape that older Millennials started.

Millennials are unlike older generations (Howe & Strauss, 2000; Poindexter, 2012). They are the most educated and most racially and ethnically diverse and they have come of age in a social media, mobile world that is nothing like the world that Generation X, Baby Boomers, and older generations grew up in. Older Millennials were in college when Facebook was born on a college campus in 2004, and while some younger Millennials choose Twitter, Instagram, and Snapchat over Facebook, older and younger

Millennials are the same when it comes to having mobile-first DNA. This Millennial affinity for mobile has enormous implications—good and bad—for the future of journalism and news engagement.

Baby Boomers may be able to afford any mobile device they choose but when it comes to mobile, they are significantly more likely than younger generations to be mobile laggards, primarily using mobile devices for communication. In a digital world, the definition of communication has expanded beyond traditional phone calls and email to include texting, messaging, and voice chat applications such as FaceTime that make it possible to see the person being talked to.

In addition to generation, race and ethnicity matter for consumers who are mobile-first and mobile laggards. Latinos and Asian Americans are more likely to be mobile-first while whites and African Americans are more likely to be mobile laggards. Other variables that in the past correlated with news consumer behaviors, attitudes, and social norms are irrelevant to being mobile-first or a mobile laggard. Women are equally as likely as men to be mobile-first, and mobile-first consumers are just as likely to be found among the highly educated and high-income as less educated and poorer consumers. Furthermore, political affiliation, including Tea Party membership, and ideology do not distinguish mobile-first consumers. Mobile-first consumers are distributed equally among Republicans, Democrats, and Independents, as well as conservatives, moderates, and liberals.

Mobile has not just transformed the consumer and media landscape, it has, in many ways, taken the traditional narrative about race, ethnicity, generation, gender, income, education, and political affiliations and turned it upside down so that old rules and conventions no longer apply and traditional gender differences, income and education inequalities, and political divides have disappeared.

Mobile Specialists

While mobile-first consumers use their devices for everything and mobile laggards limit their devices to communicating, one-third of mobile device owners mostly use smartphones and tablets to engage in specific activities. These mobile specialists, according to Table 1.1, include social media devotees (19%), Internet hunters (18%), gamers (12%), entertainment fans (9%), and news seekers (8%). Other specialists in the mobile consumer landscape have been combined into one single eclectics category, representing one-third of all mobile specialists.

Table 1.1: Specialists in the Mobile Landscape.

Type of Mobile Specialists	Percentage
Social media devotees	19
Internet hunters	18
Gamers	12
Entertainment fans	9
News seekers	8
Eclectics	33

Social Media Devotees

Devotees of social media have been around since social networking sites first emerged and could be accessed on desktops and laptops. But in today's mobile world where apps facilitate instantaneous connection to Facebook, Twitter, LinkedIn, Instagram, and Snapchat from wherever consumers may be, social media have expanded into and become intertwined in almost every aspect of consumers' lives.

Social media did not begin with Facebook, yet in a little more than a decade, the social media giant has become the standard that other social networking sites are judged by. When, in 2004, Facebook was launched from a desktop computer in a Harvard University dorm room, the social networking site was restricted to college students; today, Facebook is a publicly traded multi-billion-dollar tech corporation available to anyone 13 years or older. In 2014, Facebook had $12.47 billion in revenues, 1.39 billion monthly active users, and 1.19 billion of those monthly active users connected to Facebook using a mobile device (Facebook reports, 2015).

"Hatched" two years after Facebook was launched, Twitter, with its maximum 140-character tweets, has also grown into a multi-national public company used around the world. With 230 million active mobile users monthly, Twitter may be one-fifth the size of Facebook, but in the mobile space where news and social media intersect, Twitter plays a major role (Twitter, 2015). As a platform for real-time news reporting, real-time news engagement, and even as a real-time source for journalists, Twitter is more than a social networking site for social media devotees. Twitter is a news reporting platform and a source for news that is not just transforming how consumers engage with news, it is both competing with and driving consumers to legacy news sites.

And since Occupy Wall Street and the Arab Spring, Twitter has also become a potent tool for social activists (Murthy, 2013). Adding a hashtag with a descriptor of the movement to a tweet plus a photograph or graphic has become particularly effective in organizing and finding tweets about the topic for social activist leaders and followers and even the news media. For example, while #BlackLivesMatter was founded by three black women after the acquittal of George Zimmerman in the shooting death of Trayvon Martin, an unarmed black 17-year old in Florida, it was the hashtag that helped focus national attention on the shooting death of Michael Brown, an unarmed black teenager shot and killed by a white police officer in Ferguson, Missouri (A herstory of #BlackLivesMatter, n.d.). In fact, the #BlackLives-Matter hashtag became so powerful in focusing national attention on the deaths of unarmed black men and boys at the hands of white police officers in Ferguson, New York, North Charleston, South Carolina, Baltimore, Cleveland, Chicago, and beyond that it was among Twitter's top news hashtags of 2015 (Ortiz, 2015).

Twitter has also become an effective promotional tool for companies, organizations, sports teams, journalists, and politicians, especially Democratic and Republican presidential hopefuls in the 2016 presidential election. Presidential candidates tweet real-time comments and journalists use the tweets to both report news about the presidential election and provide context and opinion from rival candidates.

Additionally, Twitter has evolved into a platform for expressing public opinion about current events in the U.S. and around the globe. Using the hashtag #PrayForParis, the world mourned with Paris after ISIS brutally murdered 130 people there on November 13, 2015 (Ortiz, 2015). And when the U.S. Supreme Court ruled that same-sex marriage was legal on June 26, 2015, supporters of the landmark ruling used #LoveWins to celebrate (Ortiz, 2015).

LinkedIn may be a social networking site but it is about the professional—not the social. With more than 400 million members worldwide, including executives from every Fortune 500 company, LinkedIn is the "world's largest professional network on the Internet" and its stated mission is to make professionals "more productive and successful" (LinkedIn, n.d.).

While the mission of some social networking sites is to be social or professional first, some social media thrive to be visual first, which is easy to do because of the excellent camera and editing functions built into mobile devices. Instagram, which was bought by Facebook in 2012, tells users to "snap a photo with your mobile phone, then choose a filter to transform the image into a memory to keep around forever" (Instagram, n.d.). Since altering photographs violates journalism's ethics, the filters that make Instagram popular can restrict the social media site's potential as a journalistic tool. Still, Instagram is a potential source for photos, especially when news breaks such as on a snowy March day in 2015 when Delta Flight 1086, carrying 127 passengers en route to New York's LaGuardia Airport from Atlanta, skidded off the runway upon landing, smashed through a fence and "came to a rest on a snowy embankment only a few feet from the frigid waters of Flushing Bay." *USA Today* illustrated its coverage with two photos posted on Instagram (Stanglin & Mutzabaugh, 2015). Recently, Instagram added video so it is now also a video-sharing site.

Pinterest, which was founded in 2009, describes itself as a "visual bookmarking tool that helps you discover and save creative ideas" (Pinterest, n.d.) and Snapchat, which was founded in 2011 by three Stanford University students, first caught the attention of high school students. The *New York Times* said, "Snapchat made its name by enticing users with the ability to send disappearing photo, video and text messages from their smartphones. A typical message lasts no more than 10 seconds; users can send messages to many people at once" (Isaac, 2015). Snapchat is now expanding from "disappearing mobile photo sharing" (Snapchat, n.d.) into journalism (Lichterman, 2015).

None of these social media sites were established to play a role in the news media landscape, but they are playing a role nonetheless. Social media are frequent subjects of technology news stories, sources for news, and platforms for reporting or accessing news. In many ways, social media's role in the news media landscape is a good thing but it can also be bad because the "unprecedented anytime, anywhere access to social media through mobile devices" can be a "potentially powerful impediment to engaging with news, especially for Millennials (Poindexter, 2012, p. 18).

Internet Hunters

At 18%, there are about as many mobile specialists who are Internet hunters as there are social media devotees. Mobile Internet hunting, which can be deliberate and strategic or spontaneous and unplanned, is easy, convenient, educational, and even fun. With a search app such as Google and touch of a fingertip, Internet hunters can explore the limitless Web, hunt for hidden treasures, discover trending topics, and look stuff up on a whim anytime and from anywhere.

Gamers

There are no rules for Internet hunting but there are rules for playing games, which is what 12% of mobile specialists primarily do with their mobile devices. Mobile gamers play games to compete, share and bond, have fun, and pass time. Mobile games did not start with smartphones and tablets; they first emerged on a single-function handheld device called Game Boy in 1989 when Millennials were starting first grade. Launched by the Japanese company Nintendo, Game Boy was "originally bundled with Tetris," which, according to the company's website, made the handheld game-playing device an "instant phenomenon" (Nintendo, n.d.). And Game Boy secured Tetris's place in game history as the first mobile game.

Just as Tetris was bundled with Game Boy, games are bundled with mobile devices because any game can be played by downloading an app or connecting through a browser. There are apps for board games and card games such as Monopoly, Scrabble, Bingo, Solitaire, Poker, Blackjack, and Gin Rummy and there are game apps such as Candy Crush, Words with Friends, and Minecraft created to play on mobile devices. Mobile games typically have three price structures: completely free; freemium or free to download with in-app purchases; low priced from 99 cents to premium prices as high as $349.99 (Nations, n.d.; Most Expensive, n.d.; Valdez, 2015). Tetris, which was born on Game Boy as the first mobile game in the 20th century, was re-born in the 21st century to be played on smart mobile devices with two-tier pricing: freemium or 99 cents to download.

Entertainment Fans

Just as gamers may pay to play games on their mobile devices, entertainment fans often pay downloading and streaming subscription fees for TV shows, movies, and music. While it would be impossible to make the case that watching a TV program or movie on a smartphone or tablet screen is equivalent to watching it on a flat screen TV or at a movie theater, a case could be made that using earbuds or headphones to listen to music on a smartphone is the best way to listen to music outside of a music studio. Listening to music on a smartphone is just that good! Regardless of whether the viewing or listening experience is better or worse than conventional platforms, the quality of the experience may be less of a motivator for entertainment fans than mobile devices' anytime, anywhere access.

The history of listening to music with headphones plugged into a mobile device begins almost three decades before the introduction of the iPhone. It was 1979 and Generation X was just growing into its teenage years. The mobile device was called The Walkman and it was launched by Sony, the Japanese electronics and entertainment conglomerate. This device that played music on the go was so successful that *Time* magazine included it on its list of 10 "Most Influential Gadgets and Gizmos" (Most influential, n.d.)

Almost a quarter of a century after the introduction of The Walkman, the next big mobile music thing arrived and it, too, was included on *Time* magazine's list of "Most Influential Gadgets and Gizmos." By now Generation X was in its mid-30s and the first Millennials were going to college and no one had heard of Facebook yet because the social networking site was still three years in the future. It was 2001 and the mobile music gadget that Apple launched was called an iPod. The single-function mobile music device not only stored and played music, it also was small enough to fit in a pocket (Most Influential, n.d.).

Six years later, Apple bundled the iPod with its new iPhone, which not just transformed the consumer and music landscape but also cultivated a mobile specialist who was a fan of entertainment that included music, TV, movies, and new entertainment offerings such as live events, videos, and podcasts.

Fans of entertainment, of course, are not new and neither is the important role entertainment plays in society and communication. In fact, sociologist Charles Wright elevated the importance of entertainment, which he defined as "communicative acts primarily intended for amusement," by adding it to a three-activity model of communication proposed by one of the leading political scientists of the 1940s (Wright, 1975, p. 9). The three other activities were surveillance, correlation, and transmission of social heritage.

Entertainment has also played an important role in uses and gratifications, a theory that addresses communication from the perspective of the audience or consumers rather than the communicator. Entertainment both motivates and satisfies consumer needs. Every generation has had its entertainment favorites and fans, and generations dating back to the beginning of the 20th century have experienced entertainment interwoven with media and technology. The Greatest Generation, born 1901 to 1924, and the Silent Generation, born 1925 to 1945, grew up as fans of movies and comic books. The Baby Boomer Generation, born 1946 to 1964, was also fans of movies and comic books, but when TV came on the scene, Boomers became its biggest fans, which is why there was such a rapid adoption of TV in Baby Boomer homes.

Favorite TV shows included *The Ed Sullivan Show* and *American Bandstand*. As Generation X, born between 1965 and 1982, grew up, cable's MTV, which was launched in 1981, was becoming popular and movie watching expanded beyond movie theaters, thanks to VCRs, which launched a new video rental industry that made it possible to rent movies and watch them at home at one's convenience (MTV, n.d.). Today the Millennial Generation, with access to mobile devices and the development of streaming technology, has both redefined and expanded how, when, and where TV shows and movies are watched and music is listened to (Graham, 2015).

Hulu and Netflix started streaming content in 2007 and they were joined by Amazon in 2011. By 2015, HBO, Apple, Sony, and Dish launched streaming services, giving entertainment fans even more TV and movies to watch on their mobile devices (Steel, 2015).

News Seekers

In 1964, two journalism researchers proclaimed newspaper reading to be "one of the most thoroughly institutionalized behaviors of Americans" because 86% of the population read a daily newspaper (Westley & Severin, 1964, p. 45); over a half century later, despite mobile devices increasing access to news anytime and from anywhere, engagement with news is anything but an institutionalized behavior. In fact, there was such a concern about the decline in the public's engagement with news that in 2014, the Association for Education in Journalism and Mass Communication (AEJMC, n.d.) sponsored the first-ever News Engagement Day to make engaging with news a priority again. This association of journalism and communication educators, media professionals, and graduate students was right to emphasize the importance of engaging with news because in the mobile world we live in, only 8% of mobile specialists use their mobile devices primarily for news.

Eclectic Specialists

While two-thirds of mobile specialists use their mobile devices for social media, Internet hunting, playing games, accessing entertainment and news, one-third of specialists use their mobile devices for a range of other activities, including shopping, banking, reading books, online classes and school work, and GPS. These other specialized activities were gathered into one group and labeled "eclectics" in Table 1.1.

Gateway or Barrier to News?

Getting or seeking news for news seekers does not rise to the 86% of a half century ago when reading newspapers was called a "thoroughly institutionalized behavior," yet it can be seen in Table 1.2 that mobile specialists who are news seekers are more likely than other mobile consumers to get news and seek news on a daily basis. In fact, 7 out of 10 (70%) news seekers who use their mobile devices primarily for news, seek news seven days a week. But for mobile-first consumers, mobile laggards, and non-news oriented mobile specialists, daily news seeking is less of a priority. Less than half of mobile-first consumers (48%) and mobile laggards (47%) seek out news every day; for mobile specialists who aren't news seekers, only 35% seek news every day.

Table 1.2: Comparing Daily News Engagement in the Mobile Landscape.*

	News Seekers	Mobile-First	Mobile Laggard	Other Mobile Specialists
Get news 7 days	73%	63%	59%	52%
Seek news 7 days	70%	48%	47%	35%

*Statistical significance of get news is <.10; statistical significance of seek news is <.01.

Whether it's passively getting news or actively seeking it, mobile devices should be ideal news-accessing platforms. But the reality is, on mobile devices, news is not a high priority. The overwhelming majority of mobile device owners primarily use their smartphones and tablets for something other than news. The reason for the low priority may be because news is losing out to the social media and entertainment offerings that are also instantaneously available on mobile devices. Perhaps the low priority can be attributed to how news is written, displayed, or prioritized on mobile devices or maybe society places less value on being informed. The public's attitudes toward journalism, journalists, and news may also contribute to news' low priority. Or maybe it's the changing of the generational guard; the Millennial Generation that is taking over is less interested in news than generations that came before it. There are many possible reasons for the low priority of news but none of them can be fully understood without first understanding today's mobile world from a mobile-first perspective.

Living in a Mobile-First World

Since mobile-first consumers use their devices for everything, it's not surprising that in their lives, screens are center stage. For consumers who are mobile-first, smartphones and tablet computers permeate work, leisure, eating, working out, household chores, shopping, banking, classroom learning, and even sleeping. Whether at the airport, the supermarket, sports arena, movie theater, in the bank, at the gym, or on a college campus, mobile devices are in purses, pockets, or backpacks or are held in hands; strapped onto arms; or rest on desks, coffee tables, night stands, and kitchen counters. Mobile devices accompany owners to the laundry room, office, park, church, bathroom, and bed. They travel with their owners in cars and trucks; on planes, trains, and buses; and even on bikes and motorcycles. Mobile devices go on vacation; they are in hotel rooms, on beaches and boats, and poolside. They are at the restaurant, the golf course, and tennis court. They attend meetings, go to conferences, and are present for doctors', dentists', hair appointments and even manicures and pedicures. In a mobile-first world, consumers are not just enamored of their devices, they are dependent on them. Like keys, photo IDs, and debit cards, it's rare to leave home without one's mobile device. But just because mobile devices are always with their owners, it doesn't mean their owners are using them to access news.

From local news to news from distant places, accessing news with a mobile device any day or time is effortless and immediate. There's no waiting for the local or network news broadcast. There's no need to subscribe to a cable or satellite company. And there's no need to wait for the paper to be thrown on your driveway at 4 a.m. To get news, you don't need to be in your car with the radio on. Despite the fact that many news sites require a digital subscription after reading a certain number of articles for free, you don't have to have a subscription and you don't have to have a desktop or laptop computer to get news. To access news today, all you need is a mobile device. But that same mobile device that makes anytime, anywhere accessing news so easy is the same device that makes it equally as easy to access everything that is *not* news. The mobile device is not just a gateway to non-news and news offerings; it's also a marketplace where non-news industries and entrepreneurs aggressively compete with journalists and news organizations for the fingertips and screens of mobile consumers.

Winning the Fingertips of Mobile Consumers

News generally loses out to non-news offerings on smartphones and tablets. Whether mobile-first, mobile laggards, or non-news mobile specialists, news is not a priority. Any hope of elevating the importance of news on mobile devices must begin with understanding news consumers before and after smart mobile devices became a platform for delivering news. Clicks and metrics may tell from where consumers came to a mobile news site, how long they stayed, and what they did, but only the variables and factors displayed in Figure 1.2 can provide insight into why some mobile device owners seek and engage with news on their smartphones and tablets but others don't.

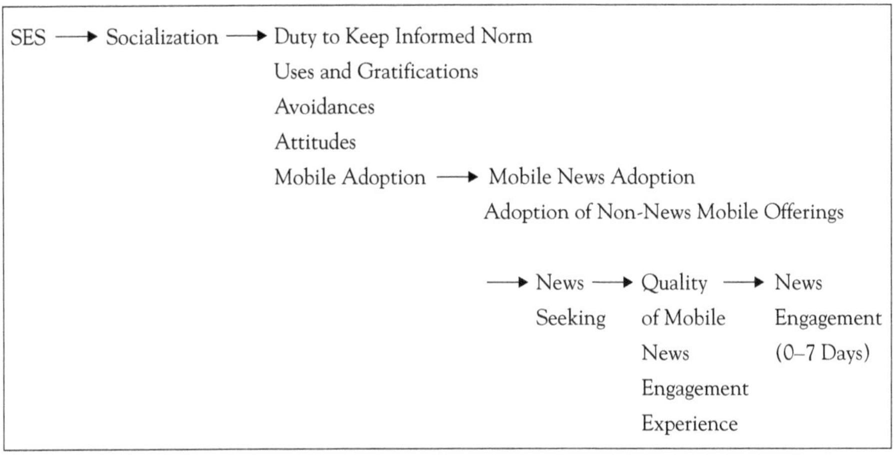

Figure 1.2: Factors Facilitating and Impeding News Seeking and Engagement in the Mobile Landscape.

According to Figure 1.2, whether or not there is news engagement on a mobile device begins with the mobile device owner's individual background or SES, which can influence how the mobile device owner is socialized about norms, uses and gratifications, and attitudes related to both mobile and news. These variables can then influence adoption of mobile devices and the use of mobile devices for news and non-news offerings, such as social media, games, and entertainment. When news is accessed on smartphones and tablets, the quality of the mobile news experience can also play a role in whether mobile device owners continue to use their smartphones and tablets to engage with news. Because news engagement on smart mobile devices is about a decade old, it is important to examine the more than 60 years of research literature on news consumers to determine whether or not there's data that may provide insight about consumers, despite their very different mobile world.

SES, or socioeconomic status, which includes demographic variables such as level of education, income, age, gender, and race and ethnicity, has been analyzed more than any other factor in news consumption behavior. Research has consistently found that educated and high income adults are more likely than adults with less education and income to engage with news. Research dating back to the 1960s and 1970s, as well as contemporary research, has also found a consistent relationship between age and news consumption. Young adults are less likely than older adults to engage with news.

The generational cohort that one belongs to has also been linked to news engagement and news avoidance. Studies have shown that the Millennial Generation is not only less likely than older generations to engage with news but it's conceivable that, unlike older generations, Millennials may not engage with news once they mature. In 2012 I identified several reasons that Millennials may be less committed to news than older generations, and one of the culprits I identified was social media, which emerged at the same time this generation was growing into adulthood. "Social media not only eats up time; it can consume attention and render traditional socializing agents ineffective when it comes to engagement with news" (Poindexter, 2012, p. 18).

In addition to age and generational cohort differences, some research has identified a gender gap in news consumption, with females consuming news less than males. A racial and ethnic divide has also been found in news consumption research, with whites engaging with news more than blacks and Latinos. That difference, though, may be more of a reflection of the higher level of education of whites than people of color. While research has identified SES variables related to news consumption, the most important value of SES may be the role it plays in socialization.

Socialization is a life-long process in which individuals acquire knowledge, values, beliefs, attitudes, behaviors, and cultural norms, and learn society's rules from socialization agents such as parents; teachers; media; peers; reference groups; marketers; local, state, and federal governments; religious and political institutions; corporations; and non-profit organizations. Of the various agents of socialization, parents are the first and most powerful because their influence begins at birth. And it's the SES of parents that influences the knowledge, values, beliefs, attitudes, behaviors, and cultural norms emphasized during the socialization process.

Parents also act as socialization agents for their children's acquisition of news engagement behaviors and attitudes. Historically, parents who read newspapers or watched TV news were role models for their children and even if the message wasn't explicit, children still learned that news was important because they observed their parents reading newspapers and watching TV news. In fact, 61% of Millennials and 84% of Baby Boomers recalled that while growing up "newspapers were around home most or all of the time" (Poindexter, 2012, p. 38). And 42% of Millennials and 52% of Baby Boomers said while growing up TV was "tuned to news around the dinner hour most or all of the time" (Poindexter, 2012, p. 39). These memories of the presence of newspapers and TV news in the home no doubt influenced later news consumption behavior. In fact, a recently published study documented that link. Using data from the University of Michigan's Panel Study of Income Dynamics (PSID), journalism scholars tested the socialization effects of parental news consumption and found "parent newspaper reading and news discussions with youth play roles in the development of news behavior among kids, in both the short-term and long-term, and especially among kids living in high SES households" (York & Scholl, 2015, p. 694).

Teachers have also been influential socialization agents when it comes to news. Two-thirds (67%) of Millennials and 55% of Baby Boomers recalled teachers including news in the classroom at least once weekly. Personal experience is not a socialization agent in the technical sense; still personal experience can perform a similar function as a socialization agent because a positive or negative experience can influence attitudes, behaviors, and the adoption of innovations relevant to engagement with news on mobile devices.

Importance of Being Informed Norm

"Believing that it is important to be informed is not a guarantee that people will actually inform themselves; still, if being informed isn't valued, it's unlikely that engagement with news will be a high priority" (Poindexter, 2012, p. 25). Because "the impact of social status is mediated through socialization" and level of education correlates with the societal norm that emphasizes the importance of being informed, individuals growing up in higher educated homes are more likely to have been taught that it's important to be informed (Milbrath & Goel, 1977, p. 53). In fact, more than two-fifths of Baby Boomers and Millennials surveyed recalled that someone had told them being informed was important (Poindexter, 2012).

Uses and Gratifications

While media consumers have always had a mind of their own, they have not always been viewed that way. In fact, for a long time the prevailing belief by political scientists and communication researchers was that communication messages were all-powerful and could penetrate and influence consumers. As experts began realizing consumers were "active" and independent, uses and gratifications theory emerged to help explain the consumer perspective. Sometimes the audience used media but it wasn't necessarily as intended and sometimes the audience outright ignored media. Uses and gratifications tried to identify what motivated consumers and the gratifications or benefits received from the media content consumed. Uses and gratifications also makes it clear that getting informed or what this approach calls surveillance is not the only motivation for engaging with news. Other motivations for engaging with news have included entertainment, escape, a tool for daily living, social contact, stimulation, habit, killing time, and social prestige (Berelson, 1949; Kimball, 1959; Katz et al., 1974; Severin & Tankard, 2001).

A survey following the shocking 9/11 terrorist attacks that killed more than 3,000 people in the U.S. found that curiosity and fear, as well as anxiety, should also be added to the list of reasons for consuming news (Poindexter & Conway, 2005). That survey also learned that the surveillance dimension of uses and gratifications was more nuanced. While most of the survey respondents paid attention to news about 9/11 to keep up with the latest developments, other examples of surveillance were: "keeping up because of personal impact (8 percent), looking to the news to learn about new threats (6 percent), trying to ascertain the world's reaction (5 percent) and gathering information about new regulations (3 percent)" (Poindexter & Conway, 2005, p. 180).

Avoidances

Even though less attention has been paid to the reasons consumers avoid news, it does not mean news avoidance research is not important. In fact, in today's mobile landscape, avoidances are even more important because despite the fact that some of the important reasons for avoiding news may no longer be relevant, news avoiders are still among us. A synthesis of a quarter of a century of research, primarily conducted with non-newspaper readers and non-TV news viewers, has identified four categories of news avoidances that explain why consumers ignore news: (1) Rejection of medium or news, such as "dislike content, format, structure, presentation, unattractiveness, purposelessness, and amount of detail; dislike bias, distrust; disinterested; dislike reading"; (2) Real constraints, including "time deficit, time conflict, health problems, language barriers, cost, circulation problems, not proficient using Internet, do not use Internet"; (3) Perceived constraints such as "lack of time, time consuming, lack of usefulness"; (4) Other medium preferred or used such as "TV, Internet, magazines, radio" (Poindexter, 2008d, p. 40).

If these avoidances are applied to the mobile landscape, rejection of the format, structure, and presentation of news on a mobile platform as well as the real constraint of the cost of a digital subscription may be most relevant as to why some consumers ignore news on smartphones and tablets.

Attitudes

Attitudes, which are learned through the socialization process and personal experience, are a favorable or unfavorable evaluation of an object. In the case of news on a smartphone or tablet, consumers may evaluate a range of dimensions, from the brand of news to its presentation. And because on a mobile device, consumers may access news from an app, social media, web browser, or search engine, there are many things to evaluate. Additionally, consumers' attitudes about the press as an institution, journalists, and news coverage have to be factored in when talking about favorable and unfavorable evaluations of news on smartphones and tablets. Favorable attitudes can facilitate news engagement; unfavorable attitudes can impede it.

Innovation Adoption

The theory about adopting an innovation such as a new idea, practice, or device lies beneath the mobile transformation of our lives. While mobile devices have been adopted at an unprecedented pace, the same cannot be said for adopting mobile devices to access news. The vast majority of mobile device owners use their smartphones and tablets to communicate and do non-news things. Even though consumers have been slow to adopt their mobile devices to access news, adoption theory is included in Figure 1.2, because it needs to be understood so news can compete for the screens and fingertips of mobile device owners.

Adopting a new idea, device, process, or behavior is a function of five characteristics of the innovation: relative advantage; compatibility; complexity, trialability, and observability (Rogers, 2003). If these characteristics are applied to the iPhone and similar Android mobile devices, it's clear why they were adopted so quickly. By bundling so many functions into a single sleek device that fits in a pocket, the iPhone did not just have a relative advantage, it was superior to pre-iPhone phones. Plus, it was easy to use and the innovative Apple stores facilitated trialability. Additionally, observability was at a high level; if someone around you had an iPhone, you just couldn't take your eyes off it and iPhone owners loved showing off the phone's cool features.

Adopting news on a mobile device is not the same as adopting a mobile device even though both fall within the parameters of adopting an innovation. The relative advantage of a smartphone over a landline or cellular phone is clear but for consumers accustomed to getting news from TV, their laptop, car radio, or even a print newspaper, the relative advantage of a mobile device is less clear, especially because for many, getting news is not just a habit, it's a ritual that may include a cup of coffee or tea. And because a smartphone demands both your attention and fingertips, it's not that easy to enjoy a cup of coffee or tea with your news, which for some puts news delivered on a mobile device at a competitive disadvantage.

As far as adopting a mobile device to access news, mobile devices also score low on trialability and observability, important characteristics in the adoption process. How many prospective mobile device owners experiment with accessing news when deciding to buy a smartphone or tablet? And observing someone accessing news on a small screen device isn't easy and may even be perceived as invading one's privacy, because unlike reading a newspaper or watching TV news, which are public and communal activities, activity on a mobile device is generally private unless the owner grants permission to observe.

In addition to the five characteristics relevant to adoption of an innovation, Rogers (2003) identifies five adopter types: (1) Innovators who are "venturesome" and "eager to try new ideas" represent 2.5% of the population; (2) early adopters who serve as a "role model" for later adopters represent 13.5%; (3) early majority who are "deliberate" in their decision to adopt, represent 34%. (4) The late majority also represents 34% and is "skeptical" about adopting an innovation. (5) Laggards are the last to adopt and they represent 16% of the population.

There are some similarities between Rogers's adopters and the mobile consumer types identified in Figure1.1. Rogers's 2.5% innovators are similar to the 2% mobile-first consumers. The 68% early and late majority are similar to mobile laggards who make up 66% of mobile device owners. And the 13.5% early adopters may come closest to the 32% of mobile specialists. Rogers's laggards would be equivalent to non-owners of mobile devices because they are not just dragging their feet, they are genuine stragglers.

Knowledge of the three types of mobile device owners identified in Figure 1.1 and awareness of the Figure 1.2 factors that contribute to news engagement are necessary to understand the factors beyond clicks and metrics to succeed in a mobile world. If news organizations can't make news more of a priority for mobile-first, mobile specialists, and mobile laggards, news engagement and the importance of news in society will continue its downward slide.

Is mobile the last chance to once again make news engagement a "thoroughly institutionalized behavior?" Maybe. Maybe not. When in 2009, the Pulitzer Prize–winning *Washington Post* editor showed off his iPhone to a journalism class, he was inserting mobile into the conversation on future challenges facing journalism. Almost a decade later, how have news organizations responded to the mobile challenge? Has it been slow and predictable or with urgency and innovation? And will the response be enough for news to become a priority on consumers' mobile devices? Will accessing news on a mobile device become as important as spending time with social media; playing games; hunting treasures on the Internet; or streaming a movie, TV show, or music? *News for a Mobile-First Consumer* will answer these questions and more. A systematic examination of the mobile landscape, including news, social media, Millennials, racial and ethnic groups, women and men, political identity, and doing mobile journalism, will provide new insight about the future of news that is nothing like the past.

· 2 ·

NEWS IN THE MOBILE AGE

News in the mobile age is more different than at any time in the history of journalism; yet in many ways news is also the same, despite reaching consumers on different platforms using different delivery systems and styles. Whether news is reported in a newspaper or magazine; on radio, TV, or cable; or on a desktop, laptop, smartphone, or tablet, news is still "a window on the world" that "aims to tell us what we want to know, need to know, and should know" (Tuchman, 1978, p. 1). Unfortunately, the public's confidence in the press's ability to open that window to inform us about what we want, need, and should know has declined to an all-time low.

A glance at Table 2.1, which compares Americans' confidence in 13 different institutions, shows how little confidence the public has in the press. In fact, the public's confidence in the press is so low that it is No. 12 on a list of 13 institutions (Confidence in institutions, n.d.). A measly 7% of the U.S. public has a "great deal" of confidence in the press, which is only two percentage points higher than the public's low opinion of Congress, the last institution on the list. Although no institution can claim confidence above 50%, the public's confidence in the press is a fraction of the public's confidence in the military and scientific community. Even confidence in education exceeds that of the press; the public is three times more confident in education than the news media.

Table 2.1: Confidence in the Press and Other Institutions.
% Saying "Great Deal" of Confidence

Military	50%
Scientific community	41%
Medicine	38%
Education	25%
Supreme Court	23%
Organized religion	19%
Major companies	18%
Banks	15%
Organized labor	12%
Executive branch	11%
Television	10%
Press	7%
Congress	5%

Note: The institutions and the confidence levels were compiled from Confidence in institutions: Trends in Americans' attitudes toward government, media, and business (Confidence in institutions, n.d.).

Confidence in the press has not always been this low. In 1976, when Baby Boomers were starting to turn 30, confidence in the press was at its highest at 28.3% (Trends in Public Attitudes, 2013). By 1998, when the Internet was being widely adopted as a consumer medium, confidence in the press dropped to single digits for the first time; only 9% expressed a great deal of confidence in the news media. That low confidence level remained unchanged over the next decade, and by the time mobile devices had reached near saturation and the first Millennials started turning 30, confidence in the press dropped to its lowest ever—7%.

The low level of confidence in the press has no doubt contributed to the half-century decline in news engagement but the impact doesn't stop there. A lack of confidence can also impede socializing the next generation about the importance of being informed about news. If parents and teachers, the primary socializing agents, have little confidence in the press, news engagement will be viewed as unworthy of passing on to the next generation.

One would expect that the principles and practices of professional journalism would boost the public's confidence in the press. Unfortunately, much of the public is in the dark about the principles that guide journalism. The public does not even know for sure the purpose of journalism or that a "discipline of verification" is one of 10 essential elements distinguishing journalism from unverified information that circulates in the news media landscape (Kovach & Rosenstiel, 2014). How would the public know the purpose of journalism or journalism's mandatory requirement that information be verified before it is published, broadcast, or posted? Journalism is not required in the K–12 curricula and it's not required in college for non-journalism majors. And unlike social networking sites such as Facebook and Twitter that post their operating standards, news organizations seem loath to explain journalism's purpose, principles, or process to the public. Because of journalism's lack of transparency about the principles, process, and elements of journalism, the public's understanding of journalism comes primarily from around-the-clock legitimate and not-so-legitimate news offerings, the popular sport of press bashing, and the portrayal of journalists in film and television, which is often as unethical vultures. If, for example, the public believes journalists are like the ones in the 2014 Academy Award–nominated *Nightcrawler*, which portrays local TV news journalists as unethical and even criminal, it's understandable why journalism has such a low standing in the public.

The news media's critiques of itself, especially when journalism does something wrong, also sour the public on the press. These critiques, which take journalism to task for plagiarism, inaccuracies, fabrication, exaggeration, and bias, are regularly found in reports from ombudsmen and Sunday morning programs such as CNN's *Reliable Sources* and Fox's *Media Buzz*. And when there is a media scandal such as the retracted *Rolling Stone* magazine story about a gang rape of a young woman at a University of Virginia fraternity, all news organizations make it the top news story. Even though the story started to unravel shortly after it was published, it wasn't until an independent investigation by Columbia University's Graduate School of Journalism was published that it was learned that basic fundamental reporting practices in terms of multiple sources, verification, and fact-checking were not followed by the reporter. This "autopsy" of the published story received as much, if not more, attention than the original story. The University of Virginia president's criticism of the retracted story, which was published in the *Washington Post* and other major news outlets, gave voice to the low level of confidence the public has in the press:

> *Rolling Stone*'s story, "A Rape on Campus," did nothing to combat sexual violence, and it damaged serious efforts to address the issue. Irresponsible journalism unjustly damaged the reputations of many innocent individuals and the University of

> Virginia. *Rolling Stone* falsely accused some University of Virginia students of heinous, criminal acts, and falsely depicted others as indifferent to the suffering of their classmate. The story portrayed University staff members as manipulative and callous toward victims of sexual assault. Such false depictions reinforce the reluctance sexual assault victims already feel about reporting their experience, lest they be doubted or ignored. (President Teresa A. Sullivan Statement, 2015)

The retracted gang rape story was not just poor reporting by the reporter; it was a professional failure of what distinguishes journalism from other information that circulates as journalism: verification. The damage from *Rolling Stone*'s unverified rape story was not limited to the magazine; the journalism profession was also damaged. And despite the fact the journalism malpractice like the retracted *Rolling Stone* fraternity gang rape article is rare, news reports on these infractions usually fail to mention that fact. And without that context, the public's low opinion of the press is reinforced as confidence declines further.

While the First Amendment to the U.S. Constitution is sacred to the press and U.S. citizens, to screenwriters, the First Amendment is often a plot point where the press is often portrayed as hiding behind the First Amendment to protect a source even if the source is a terrorist or has committed a crime. Without the press helping the public understand that First Amendment protections are essential to providing them with the "information they need to be free and self-governing," plot point TV and film portrayals of the First Amendment will define how the press operates, which will more likely than not contribute to the public's low estimation of the press (Kovach & Rosenstiel, 2014, p. 17).

And the news media's history of excluding and stereotyping people of color and women, and most recently Millennials, has also contributed to the public's low confidence because it calls into question journalism's ability to deliver on one of five requirements the 1947 Hutchins Commission report said "society is entitled to demand of its press" (A free and responsible press, 1947, p. 26). This requirement—"projecting a representative picture of the constituent groups in the society"—(p. 26) speaks directly to news coverage of people of color, women, and Millennials who traditionally have been excluded as subjects and sources, stereotyped, and treated as "second-tier" (Poindexter, 2008b, p. 72).

Two decades after the Hutchins Commission report declared the press has a responsibility to project a representative picture, the Kerner Commission criticized the press for its failure to include blacks in that picture (Report of the National Advisory Commission on Civil Disorders, 1968). The landmark report that examined the reasons behind the 1967 riots across America also criticized newsrooms for failing to diversify, an issue that is still of concern today. In the U.S., African Americans, Latinos, Asian Americans, and Native Americans make up 37% of the population but only 12.76% of the 32,900 newsroom staff at 1,400 daily newspapers are racial and ethnic minorities and in TV newsrooms, the percentage of racial and ethnic minorities is 22.2% (Percentages of Minorities, 2015; Papper, 2015). The news media's long-standing failure to achieve fully diverse newsrooms has no doubt contributed to diversity shortcomings in news topics, subjects, sources, and framing of the news, regardless of platform delivering the news. Furthermore, this failure to diversify likely diminishes the confidence in the press of the very people who have been excluded from and stereotyped in the news.

The press has also failed to explain to the public that certain criteria—audience, impact, proximity, timeliness, prominence, unusualness, conflict, visual and audio impact—have traditionally been used to determine whether or not something is newsworthy. Not knowing why certain editorial decisions are made may also contribute to the public's low level of confidence in the press. But because the criteria for newsworthiness has not always been applied with an eye toward fairness and balance, there may be an unintended consequence that people of color are stereotyped in negative ways. In fact, a review of 30 years of journalism and communication studies found that African Americans were not only more likely to be excluded from the news but if they were included, they were stereotyped in the news as criminals, poor, athletes, and celebrities (Poindexter, 2011). Latinos, Native Americans, and Asian Americans were far more likely than African Americans to be excluded from the news but when they were included, they too were stereotyped.

In the age of YouTube, Twitter, and smartphones, "viral" and "trending" have been added to traditional criteria for newsworthiness, but reporting on what's trending may also contribute to the public's lack of confidence in the press, especially if what is reported on is perceived as trivial as whether the color of a dress is blue or black (Netburn, 2015). While unusualness may have been the reason that so much space and time were devoted to #thedress, the fact remains that from the perspective of the public, surely there was more important news that deserved attention.

Another issue that may contribute to the public's low level of confidence in the press is that in the U.S., there are no educational, exam, or licensing requirements to be a journalist. Contrast the lack of professional requirements to be a journalist with the educational, exam, and licensing requirements to become a physician, attorney, or pharmacist, and the public may feel justified in its low level of confidence in the press. Plus, for some, it is not reassuring that there are no national governing boards for journalism malpractice, especially because governing boards can disbar attorneys and cause physicians and pharmacists to lose their licenses.

Journalists do have ethical codes but the public doesn't know what they are, which can contribute to the public's low level of confidence in the press as an institution. While it is agreed that plagiarism, fabrication, and altering photos are unethical, the fact that journalism does not have a uniform ethical code that can be posted on news sites is problematic. A look at ethical codes of journalism associations such as ASNE, SPJ, and ONA reveals different language and different emphases. Even wire services such as the Associated Press and individual news organizations such as the *New York Times* have their individualized ethical codes. The hodgepodge of ethical codes, while indispensable for individual news associations and organizations, may contribute to more public confusion than confidence. It also doesn't help that some journalists consider some ethically questionable practices acceptable.

For example, a national survey of journalists found 58% of journalists agreed it was justifiable to use confidential business or government documents without authorization; 38% agreed badgering unwilling sources was acceptable; using personal letters and photographs without permission was not a problem for 25% of journalists; and 5% of journalists found paying for confidential information was acceptable (Willnat & Weaver, 2014). It is doubtful that some of these practices, especially the ones that cross the line of privacy and badger sources, willing or unwilling, would be condoned by the public.

And when prominent news anchors and hosts, such as NBC's Brian Williams, Fox's Bill O'Reilly, or ABC's George Stephanopoulos, wade into ethical hot water, the public's confidence in the press is further damaged (Battaglio & Halper, 2015). And when the press makes such a big deal of the ethical lapses without putting them in the context of journalism standards, the public may be more confused than informed about the principles of journalism.

With Facebook, Twitter, Snapchat, and other social networking sites taking up residence in the news media landscape, journalism has made these social networking sites integral to reporting and delivering news. While 80% of journalists agreed or strongly agreed that social media help them promote themselves and work, only 30% agreed or strongly agreed that social media "enhanced their professional credibility" (Willnat & Weaver, 2014).

Journalists see the promotion and credibility benefits of social media but does the public view social media as a trustworthy source of news? To provide insight into the public's perspective on social media, the News for a Mobile-First Consumer Survey asked about the level of trust about social media and other sources of news. Table 2.2 reveals that despite journalists' belief that social media enhances their professional credibility, the public does not view social media as a trusted source of news. In fact, the public ranks social media among the least trustworthy sources of news, with 14% trusting Facebook a lot as a source of news and only 9% trusting Twitter. As a profession, journalists do not fare much better than social media on the trustworthy scale—only 17% trust journalists a lot—but the product that local TV journalists produce is the most trustworthy of the 12 sources of news; 42% trust local TV news a lot. Local news published in the local newspaper's online and print editions is trusted by 34%, followed by network TV news (30%). It's worth noting that family and friends (28%) and Google (22%) score higher than journalists (17%) as a trusted news source.

Table 2.2: Trust of 12 Sources of News.

Source of News:	% Saying Trust A lot
Local TV news	42%
Local print & online newspaper	34%
Network TV news	30%
Family & friends	28%
Cable news	27%
National newspaper	26%
Google	22%
Journalists	17%
Facebook	14%
Government websites	13%
Twitter	9%
Bloggers	9%

With only 7% of the public having a great deal of confidence in the press and only 17% of the public having a high level of trust in journalists, it appears that almost two decades after 25 journalists observed that something was "seriously wrong" with journalism, something is still "seriously wrong" (Kovach & Rosenstiel, 2014, p. 2). In fact, 60% of journalists surveyed for "The American Journalist in the Digital Age" said "journalism is going in the wrong direction" (Willnat & Weaver, 2014).

Ironically, despite the public's record low level of confidence in the press and journalism going in the wrong direction, journalism has survived the Great Recession of 2008, the digitizing of the news landscape, and collapse of an extremely profitable business model. But is journalism nimble enough to catch up and thrive in today's news landscape where, without warning and in record time, smart mobile devices have taken over? Can news organizations get ahead of this mobile tsunami and thrive or will they try to tread water and get left behind by consumers who have embraced smart mobile devices with an enthusiasm unseen since, perhaps, the emergence of television almost 70 years ago.

Mobile device owners who now inhabit the news landscape are not a monolith and neither, according to Table 2.2, is their trust of varying sources of news. To provide insight into mobile owners' trust of news, the 12 sources were analyzed by the different mobile types introduced in Chapter 1: Mobile-first, who represent 2% of mobile device owners, use their mobile device for *everything*; mobile laggards, representing 66%, mostly use their mobile device for communication such as telephone calls, email, or texting; and mobile specialists who, at 32%, primarily use their mobile devices to connect to news or for engaging in specific activities such as social media, games, entertainment, and searching the Internet.

As Table 2.3 shows, type of mobile consumer is related to news source trust. First and foremost, mobile news specialists have a higher level of trust of the 12 sources of news than do mobile-first, non-news mobile specialists, and mobile laggards. And the majority (67%) of mobile news specialists trust local TV news more than any news source.

After local TV news, local print and online newspapers are most trusted, with 58% of mobile news specialists saying they trust this news source a lot. Network TV news and national news are tied with 52% of mobile news specialists saying they trust these two sources of national news a lot. Despite 39% of mobile news specialists trusting journalists as news sources more than the other mobile consumer types, it's not a ringing endorsement, since

Table 2.3: Mobile Types' Trust of News Sources.
% Saying Trust A lot

	Mobile-First	Mobile News Specialist	Other Mobile Specialists	Mobile Laggards
Local TV news	42%	67%	42%	41%
Local print & online newspaper	21%	58%	36%	33%
Network TV news	29%	52%	29%	28%
National news	25%	52%	28%	22%
Cable news	38%	46%	27%	24%
Google	17%	42%	27%	16%
Journalists	17%	39%	20%	13%
Family & friends	21%	36%	32%	24%
Facebook	12%	27%	19%	10%
Government websites	8%	21%	14%	9%
Twitter	8%	21%	12%	6%
Blogger	8%	18%	12%	5%

Note: Mobile specialists are divided into two groups: mobile news specialists and other mobile specialists, which includes non-news mobile specialists. Significance for local TV news and family and friends is <.10 and network TV news is not statistically significant.

journalists are 7th on a list of 12 different news sources. Still, for mobile news specialists, journalists are trusted as news sources more than family and friends (36%), Facebook (27%), government websites and Twitter (21%), and bloggers (18%).

It's important to note that while mobile laggards have adopted smartphones, they are using them in a more traditional way, that is, as a communication device. Still two-fifths (41%) of mobile laggards trust local TV news but trust of journalists is very low, with only 13% trusting journalists a lot.

Mobile-first consumers currently represent a small percentage of mobile device owners, but that is expected to increase as more people integrate smart mobile devices into their whole lives and the post-Millennial generation grows up. Among mobile-first consumers, only 17% trust journalists a lot, slightly lower than the 20% of non-news specialists.

The Mobilization of the News Landscape

Mobile devices have changed how, when, where, and if consumers engage with news. Mobile devices have sped up the 24/7 news cycle that first emerged when CNN was launched in 1980. As the first 24-hour cable news network, CNN changed the public's ability to access news from the appointed times when newspapers were delivered or local or network news was broadcast. When CNN was launched, for the first time ever, anyone with a cable subscription could get news 24 hours, 7 days a week. Some news was a repeat of news broadcast earlier in the day but it was still available by turning on the TV. Over a decade later, with newspapers' first experiments with news delivered through the Internet and family-friendly services such as AOL, news was also available from a desktop computer, 24/7. But this access to news was only possible by sitting in front of a TV or a desktop computer. And even though laptops with access to the Internet in coffeehouses such as Starbucks may have started the process of accessing news 24/7 at a location other than in front of a TV or desktop computer, smartphones brought about a revolution in accessing news. With smartphones, consumers could access news from anywhere and at any time of the day or night (Yang, 2010).

Despite having taken approximately four centuries to go from news that could be read in a newspaper to news available to read, watch, listen to, and share using a mobile device, this news landscape transformation can be thought of as three distinct periods as shown in Table 2.4: Period 1, Pre-Online News, beginning in the early 1600s, when the first newspapers were published according to news historian Mitchell Stephens (2007) and lasting through the early 1990s; Period 2, Online News, 1994 to 2006; Period 3, Mobile News, 2007 to Present.

The invention of moveable type led to the printing of the Gutenberg Bible in the mid-1400s, but it wasn't until the early 1600s before the first newspapers were printed in Europe (Stephens, 2007). For the next 300 years, newspapers were the only platform delivering news directly to an audience but that changed in the 1920s with the emergence of radio and news magazines. In 1920, the first commercial radio station, Pittsburgh's KDKA, was launched by Westinghouse and reported the results of the Harding-Cox presidential election, and in 1923, the weekly news magazine *Time* was founded by Henry Luce (Stephens, 2007). Radio, with its entertainment programs and news reports became the center of households but eventually gave way to TV, which was rapidly adopted.

TV news was broadcast at appointed times by the big three networks CBS, NBC, and ABC but that changed with the creation of CNN in 1980 (Stephens, 2007). For the first time, news was available 24 hours a day, seven days a week. And even if news was repeated throughout the day and night, it was available whenever consumers wanted to watch. Thanks to the creation of ARPANET, the forerunner to the Internet, in 1969, Period 1 gave way to Period 2, when accessing news expanded beyond newspapers, radio, TV, and cable (Isaacson, 2014). By the 1990s, news could be accessed online. For this online news period to take hold, personal computers had to have been invented and adopted by consumers and the Internet had to have become consumer friendly, which it did with Tim Berners-Lee's 1991 creation of the World Wide Web and Steve Case's development of AOL that made access to the Internet so easy that even kids could go online (Isaacson, 2014).

With the consumer-friendly capability of the Web, the *Chicago Tribune*, *San Jose Mercury News*, and the *New York Times* began offering limited news online and by 1994, 20 U.S. newspapers were online (Weiss, 2008). Over the next decade online news that could be accessed with a personal computer became the norm. Multimedia made it possible for legacy print news media to report the news using video and audio traditionally used only in TV and radio news. And the emergence of digital made it possible to enhance news photography, graphics, and storytelling in new and engaging ways. The unveiling of Apple's iPhone led to Period 3, Mobile News, which saw the emergence of a new type of news and news consumer. The new mobile news built on the multimedia news that had become the norm on desktops and laptops and created a new way to display and access news. In the mobile landscape, consumers could still access news using browsers but now with a tap of their fingertip, they could access news from a mobile app. But just because consumers could access news on mobile apps did not mean they did.

By following the innovations across the three Periods displayed in Table 2.4, it can be seen how once an innovation emerges, it may change form to some degree but it does not disappear. For example, the 400-year-old newspaper that emerged in Period 1 and 60-year-old TV news program that emerged in Period 2 are still with us, and so are AOL and news sites, also innovations of Period 2. The social media news of Period 2 has successfully made the transformation to mobile in Period 3. So far in the mobile news Period 3, news apps are the most significant innovation because they offer the easiest, most efficient, and most convenient avenue for consumers to connect with news, anytime and from anywhere. But news apps do not have Period 3 alone; news innovations as well as non-news content and activities from Periods 1 and Period 2 also compete for the attention of mobile device owners.

Table 2.4: News in Three Periods: Pre-Online, Online, Mobile.

Period 1 Pre-Online News: Newspapers to Cable 1605–1993	Period 2 Online News: America Online to News Websites 1994–2006	Period 3 Mobile News 2007–Present
Newspapers	→	→
Magazines	→	→
Radio news	→	→
TV news	→	→
24-hour cable news	→	→
News wire	→	→
Breaking news	→	→
Opinion	→	→
News satire	→	→
Letters to the editor	→	→
News polls	→	→
News photos	→	→
Wild art slice of life photos	→	→
Graphics	→	→
	Data-driven stories	→
	AOL	→
	News sites	→
	E-replica newspaper	→
	Online-only news sites	→
	Search engines	→
	Browsers	→
	Blogs	→
	Aggregated and curated news	→
	Emailed news	→
	News notifications	→

Table 2.4: *(continued)*

Period 1 Pre-Online News: Newspapers to Cable 1605–1993	Period 2 Online News: America Online to News Websites 1994–2006	Period 3 Mobile News 2007–Present
	Online comments	→
	Citizen journalist news	→
	Citizen photojournalist photos	→
	Fact-checking news	→
	Multimedia, including video, podcasts, photo galleries, interactives, and animation	→
	Podcasts	→
	Infographics	→
	Viral news	→
	Trending news	→
	Social media news	→
		Mobile news site
		Mobile news apps
		Mobile news alerts
		Streaming news
		Big data
		News apps curated by technology and social media companies
		Wearable
		Virtual reality
		Other

Mobile News Apps

Before the invention of mobile news apps, news was accessed by reading newspapers and news magazines, watching TV, listening to radio, or clicking on news stories on desktops and laptops. With the development of smart mobile devices and apps, news could be accessed with one's fingertip. By tapping, scrolling, and swiping the screen of a device held in your hand, news could be read, watched, listened to, interacted with, and shared. As an innovation, news apps are not just a new idea, they are also a new platform for accessing news. And to use this innovation to access news on a mobile screen, a new and different behavior that combines finger tapping, scrolling, and swiping is required.

But before news can be accessed from a news app, the mobile device owner usually has to search for the app in the App store and download it. A generic search of news apps turned up more than 23,000, representing news from around the world. Some news apps are free; others require a digital subscription that could be monthly or annual or even kick in after accessing a minimum number of stories.

Apps are not the only method for accessing news on a smartphone or tablet, but they are one of the few news innovations so far to emerge during Period 3. While this news innovation makes it quick and easy to access news, it doesn't mean it's the only method used to get news on a mobile device. According to Figure 2.1, 37% of mobile device owners use apps to get news, followed by 20% who use browsers and 19% who engage with news on Facebook. Fewer than one out of ten (9%) engage using a breaking news alert, 7%, a search engine, and only 4% engage with news on their mobile device through Twitter.

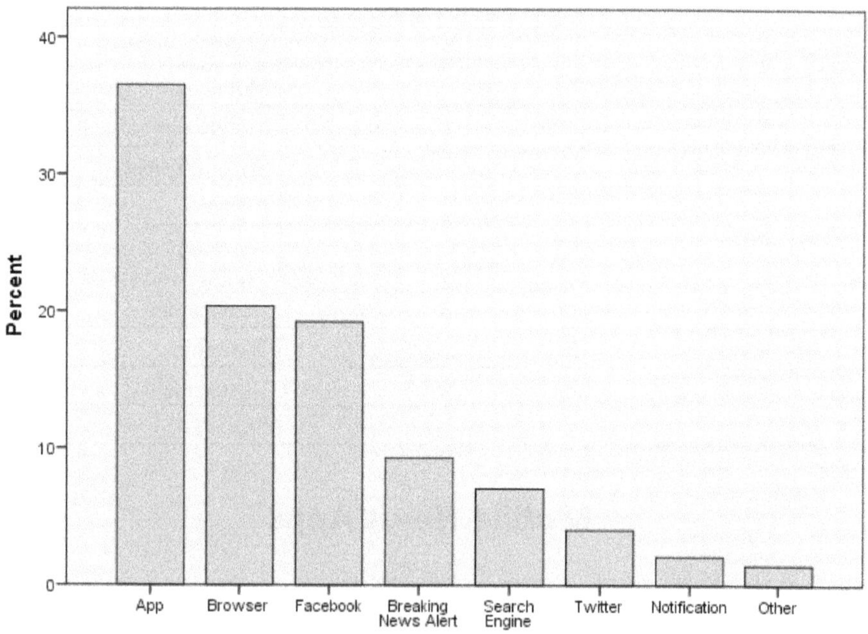

Figure 2.1: How Mobile Device Owners Engage with News.

Despite apps being the primary method for accessing news on mobile devices, mobile devices average only 2.4 news apps. And mobile laggards and non-news mobile specialists are more likely to have no mobile news apps at all. In fact, according to Table 2.5, 46% of mobile laggards and 28% of other mobile specialists have "0" mobile news apps.

Table 2.5: Number of Apps That Mobile Types Have.

	Mobile First	Mobile News Specialist	Other Mobile Specialists Excluding News	Mobile Laggards
0 news apps	4%	3%	28%	46%
1 or 2 apps	42%	36%	44%	34%
3 or more apps	54%	61%	28%	20%

The failure of mobile laggards to have news apps is not surprising because these consumers primarily use their mobile devices for communication. It's also not news that non-news mobile specialists don't have news apps. Non-news mobile specialists are interested in non-news apps, including social media, games, and entertainment. Still the paltry presence of news apps on mobile screens is of concern because without the presence of news apps on mobile screens, two of five criteria that Rogers (2003) said were necessary for adoption of an innovation are missing: trialability and observability. And without trying and observing mobile news apps, it's more difficult to know from experience and observation whether news apps have a relative advantage over browsers and other methods of accessing news. And without experimenting with mobile news apps, how can one know if news apps are easy or difficult to use or even compatible with other news platforms? Without this experience and knowledge, the adoption process is disadvantaged, which in the long run may make accessing news on a mobile device less of a priority in the lives of mobile device owners.

· 3 ·

SOCIAL MEDIA ON THE MOBILE STAGE

News apps may be absent from many smartphone screens and lists of top mobile apps, but the opposite is true for social media. One-third of the top 15 smartphone apps are for social media. With more than 1.49 billion active monthly users (Facebook, 2015), it is not surprising that Facebook is the dominant app. Facebook comprises 71% of the app audience, followed by Facebook-owned, photo-and-video-sharing Instagram, which is just over one-third (35%) of the app audience (ComScore, 2015). The 140-character Twitter and Facebook rival Google+ are less than one-fourth of mobile apps. Snapchat, one of the newest additions to the social media landscape, is 19% of the mobile app audience (see Table 3.1).

Table 3.1: Top Social Media Apps.

Social Media	Percentage of App Audience
Facebook	71%
Instagram	35%
Twitter	23%
Google+	22%
Snapchat	19%

Note: The social media apps were compiled from ComScore's Top 15 apps (ComScore, 2015).

Social media's dominance in the mobile app landscape is both good and bad for news. Social media compete with news for the attention of mobile device owners and more often than not win out. Plus, the dominant presence of social media apps, also a 21st century innovation, provides opportunities for "trialability" and "observation," characteristics required for consumer adoption. With trialability and observation opportunities, social media on mobile have the potential to become even more dominant as more and more mobile device owners adopt apps for social media but not news.

But as a platform that can be used to display and share news stories and photos, social media apps also have the potential to reach mobile device owners who may not engage with news on traditional platforms or even mobile news apps. The potential of increasing news engagement through social media is evident from social media statistics. Slightly more than three-quarters (76%) of participants in the News for a Mobile-First Consumer Survey said they were on Facebook and over three-fifths (62%) have the mobile Facebook app on their screens. More than one-third (35%) of survey participants said they were on Twitter and almost as many (31%) had Twitter's mobile app. Over a quarter (26%) of survey participants were Instagram users and 28% of Instagram users had downloaded the app on their mobile device.

Additionally, mobile social media apps offer some of the benefits of news apps. Just as news apps make it easy to access news anytime and from any place, the same is true for social media apps. But social media have one important advantage when it comes to accessing news. Technically, there's no news paywall on Facebook, Twitter, or other social media sites so there's no barrier to news for consumers without a digital news subscription. And even if a news organization limits access to the full article or number of articles that can be read in a month, there is plenty of free news on social media that consumers can find or see even if they are not seeking it.

Even so, it must be remembered that being a social media user does not automatically translate into engaging with news, even with free access. For example, 22% of Facebook users and 17% of Twitter users devote no time to news, according to the News for a Mobile-First Consumer Survey. Mobile laggards, who primarily use their mobile devices for communication, are more likely than mobile-first and mobile specialists, including mobile specialists who are social media devotees and news seekers, to avoid news on Facebook, while mobile-first consumers are more likely than the other mobile types to avoid news on Twitter. According to Table 3.2, over a quarter (27%) of mobile laggards avoid news on Facebook and over one-third (35%) of mobile-first consumers ignore news on Twitter.

Table 3.2: Mobile Consumer Types Ignoring News on Social Media.
% Who Get No News on Social Media

	Mobile-First	Mobile Laggards	Mobile Specialists Who Are Social Media Devotees	Mobile Specialists Who Are News Seekers	Other Mobile Specialists
Facebook	10%	27%	7%	11%	21%
Twitter	35%	20%	9%	0%	18%

Note: Social media devotees and news seekers are both in the mobile specialist group, but they were analyzed separately in Table 3.2.

Even though some mobile consumer types ignore news on social media, a significant percentage of Facebook and Twitter users spend at least one-third of social media time on news. Specifically, 46% of mobile specialists who are news seekers use at least one-third of their time on Twitter to engage with news and 43% spend at least one-third of their Facebook time on news. Even 31% of mobile specialists who are social media devotees use at least one-third of their Facebook time for news.

Table 3.3: Mobile Types Devoting at Least One-Third of Social Media Time to News.

	Mobile-First	Mobile Laggards	Mobile Specialists Who Are Social Media Devotees	Mobile Specialists Who Are News Seekers	Other Mobile Specialists
Facebook	26%	18%	31%	43%	26%
Twitter	29%	20%	24%	46%	34%

Note: Social media devotees and news seekers are both in the mobile specialist group, but they were analyzed separately in Table 3.3.

Social media and social media news are 21st-century innovations, but does that mean social media news consumers are similar to 20th-century legacy news consumers, who were more likely to be white, middle-aged, and educated, and beginning in the last decade of the 20th century, more male than female?

Demographics and Social Media News

Dating back to 1964 when journalism scholars found the youngest and oldest adults were less likely than middle-age adults to regularly read newspapers, study after news-consumption study has found a similar pattern. This curvilinear age pattern, though, differs for social media news consumers (Westley & Severin, 1964). According to Table 3.4, which highlights demographics (age, education, race, ethnicity, gender) and news consumption relationships in the social media environment, there is a negative relationship between age and social media news use; in other words, as age increases, social media news consumption decreases. This pattern holds for both Facebook and Twitter; 38% of the youngest adults, Millennials, compared to 11% of the oldest adults, Baby Boomers, spend at least one-third of their time on social media engaging with news. This 21st-century relationship between news via social media and age suggests the gateway to news for Millennials may be through social media. This social media gateway had already been found by the Pew Research Center for laptops and desktops. Both younger and older Millennials were more likely to get news via Facebook than a search engine or browser. The Pew Research Center found:

> In 16 out of the 19 sites where comparisons could be made, Facebook sent a significantly higher percentage of 25- to 34-year-old visitors than the direct path. And in 16 of the 17 sites that could be compared, Facebook sent higher percentages of 25- to 34-year-olds than search—often by very large margins.
>
> In an even younger age group, 18- to 24-year-olds, the same pattern emerges, though it is less pronounced. Of the 18 sites where comparisons could be made, the percentage of 18- to 24-year-olds coming from Facebook was higher than those coming directly on 15 sites and higher than the search visitors on all 18. (Mitchell, Jurkowitz, & Olmstead, 2014)

Just as the legacy news and age relationship is upended with news via social media in the 21st century, the traditional positive relationship between legacy news consumption and education is also turned upside down, at least for Facebook, according to the News for a Mobile-First Consumer Survey. For Facebook, there is a negative relationship between education and social media news consumption; 33% of adults with a high school degree or less engage with news via social media at least one-third of the time while only 21% of adults with at least a college degree spend at least one-third of their time on social media with news. The relationship for Twitter news is curvilinear; 33%

of adults with a high school degree or less and 32% of adults with a college degree or more spend at least one-third of Twitter time on news, but only 23% of adults with some college devote at least one-third of their time to news while on Twitter. Just as social media may be a gateway to news for Millennials, it may also be a gateway to news for adults who have a high school degree or less.

Table 3.4: Who Engages with News at Least One-Third of Social Media Time.

	Facebook	Twitter
Age (Generation):		
Young Adults (Millennials)	38%	38%
Middle Age (Generation X)	28%	30%
Oldest (Baby Boomers)	11%	11%
Education:		
H.S. Degree	33%	33%
Some College	24%	23%
College Degree-Plus	21%	32%
Race and Ethnicity:		
White	19%	28%
Black/African American	25%	24%
Latino/Hispanic	34%	31%
Asian American	45%	48%
Native American	13%	17%
Gender:		
Male	24%	34%
Female	26%	26%

While social media appear to facilitate news engagement for Millennials and less educated adults, social networking is flipping a pre-Internet legacy newspaper consumption relationship when whites were more likely than people of color to read newspapers. In the post-Internet, social media news environment, a smaller percentage of whites (19%) than Latinos (34%), Asian Americans (45%), and African Americans (25%) engage with news at least one-third of their time on Facebook. And on Twitter, a larger percentage of Asian Americans (48%) and Latinos (31%) than whites (28%) and blacks (24%) spend at least one-third of their time on news. For both Facebook and Twitter, fewer Native Americans engage with news one-third of the time.

The relationship between gender and news consumption has been altered somewhat in the social media environment. First, there was no gender gap in news consumption but during the final decade of the 20th century, a gender gap in news consumption emerged (Poindexter & Harp, 2008). That gender gap persisted on Twitter but not on Facebook. On Twitter, males engage with news significantly more than females (34% vs. 26%) but on Facebook, about one-quarter of males and females spend at least one-third of their time engaging with news.

News in the Mobile Social Media Landscape

News organizations' failure to replicate a profitable 20th-century business model in a 21st-century digital world has led to unprecedented threats against the future of news. And just when news organizations seemed to have figured out digital news delivered through news sites via desktops and laptops, consumers embraced mobile devices, forcing news organizations to reinvent news again, this time for smartphones and tablets. For news organizations, the transition to mobile has not been easy because of the legacy of newsprint, ink, audited circulation, Nielsen ratings, a history of lagging behind rather than leading change, and the emergence of the Millennial generation that has come of age on mobile devices and social media while ignoring legacy news as we know it. Both the dying out of last century's committed news consumers and the snubbing of legacy news by the Millennial generation led me to ask: Is news engagement a thing of the past? (Poindexter, 2012). While I theorized that social media may have been a barrier to news engagement for the Millennial Generation in the pre-mobile era; in today's mobile world, mobile social media may be legacy news organizations' best hope for engaging the youngest generation in news.

Social networking sites were not founded as platforms for news, but they have become platforms for news, nevertheless. And with the addition of mobile social media apps, news has an innovative and convenient way to connect with consumers that is nothing like the generations that connected with legacy news by reading, listening to, and watching news. Legacy newspapers, magazines, network TV, local TV, and cable and radio news have a social media presence, including Facebook pages and Twitter accounts, and they make it easy to post and share news via social media. Legacy and digital-only news embed social media logos and links and even rely on Twitter and Facebook as sources for news. News organizations are now becoming more pro-active in engaging consumers through social media by establishing partnerships with Snapchat, the latest addition to the social media terrain, and Facebook, the social media behemoth that almost overnight successfully transformed the social networking site that began on a desktop into a mobile-first platform (Blum, 2015; Lichterman, 2015).

As platforms for news, many social media sites can provide engagement opportunities above and beyond traditional legacy news reading, watching, and listening. With news via social media, consumers can post, share, "like," and comment on news, and with Facebook's introduction of reactions, users can react to the news with a variety of emotions, including sadness and anger (Krug, 2016). And with mobile social media apps and a finger tap, consumers can instantaneously engage with news anytime and from wherever they may be. Each social networking site is unique. Facebook is facilitating news consumption by serving as a gateway to news sites, especially for Millennials, and Twitter has become a Millennial go-to source for news and an indispensable reporting tool for news organizations, professional journalists, and citizen journalists. And despite the fact that 78% of Facebook users don't go to the social networking site for news, they nevertheless find news and will click on links for a variety of reasons, but still the major reasons for using Facebook are related to friends, family, sharing, and games:

> For 68% of Facebook users, seeing what friends and family are up to is a major reason they go to the site. Nearly as many, 62%, say the same of seeing friends' photos. Less commonly cited reasons include engaging in conversations with friends and family (38%) and sharing one's own photos or videos (28%). And just 16% of Facebook users say getting news is a major reason they use Facebook, on par with the proportion who say posting personal updates (17%) or playing games (14%) are major reasons. (Mitchell et al., 2013)

Whether or not one was among the 16% that news was a major reason for using Facebook, Facebook users are clear about the reasons for clicking on news. According to the results of a Pew Research Center study (Role of news on Facebook, 2013) presented in Table 3.5, the vast majority (70%) of Facebook users who clicked on news did so because of an interest in the topic. A friend's recommendation was more important than the news organization that produced it; 37% clicked on a news story because a friend recommended it but only 20% clicked on a story because of the news organization that reported it. Still, news organizations and individual journalists and commentators mattered enough for 34% of Facebook news consumers because they had proactively "liked" their Facebook pages, according to a Pew Research Center survey. However, it should be emphasized that the majority (66%) of Facebook news consumers had not "liked" a news organization, journalist, or commentator.

Table 3.5: Why Facebook Users Click on Links to News Stories.

Interest in topic	70%
Funny or entertaining story	51%
Surprising or unexpected story	50%
Follow topic in the news	45%
Story recommended by friend	37%
Story published by preferred news organization	20%
Story had lots of likes or comments	13%

Note: Compiled from Pew Research Center Report, The Role of News on Facebook: Common Yet Incidental (Role on news on Facebook, 2013).

The Twitter Effect in the News Media Landscape

Facebook was not founded as a gateway for news but that's one of the roles it has taken on. Twitter may have begun as a way to share one's status with friends but it has evolved into a multifaceted platform for reporting and engaging with news. Twitter vocabulary—following, tweeting, retweeting, trending—even has a special meaning when applied to news.

There are a variety of ways to access the social networking site, with the most popular being through Twitter's mobile app; 45% of News for a Mobile-First Consumer Survey participants reported they used the app. Almost one-third (32%) access Twitter using the browser on a laptop. Another 15% use the browser of the smartphone or tablet and 8% access Twitter in other ways.

Although the Twitter app is the most popular method for accessing Twitter, it is not the most popular social media app that mobile device owners have on their phones, according to the News for a Mobile-First Consumer Survey. At 62%, Facebook is the most popular social media app, followed by YouTube's app, which 57% of consumers have on their mobile devices. Fewer than one-third (31%) have a Twitter app; 28% have the Instagram app; 17% have LinkedIn's app; and 21% have other social media apps.

Having a mobile app appears to make a difference in engaging with news via Twitter. According to Table 3.6, mobile device owners with a Twitter app are more than twice as likely as mobile device owners without a Twitter app to follow journalists or news organizations (50% vs. 22%). When survey participants were asked which journalists or news organizations they followed on

Twitter, only a fraction mentioned specific names. Still, specific names were mentioned and they included both legacy and digital-only: Anderson Cooper, Fox, CNN, ABC, BBC, Huffington Post, NBC, *New York Times*, *Los Angeles Times*, *Wall Street Journal*, BuzzFeed, *Washington Post*, NPR, Vice, and Al Jazeera.

In addition to following journalists and news organizations, Twitter mobile app users are almost three times more likely than Twitter users without mobile apps to click on tweeted news often (26% vs. 9%). An app also seems to matter when it comes to news trending on Twitter: 25% of Twitter users with an app click on trending news more often compared to 13% without a mobile app. Retweeting news seems to be the least popular method of engaging with news on Twitter. Still, Twitter users with a mobile app retweet more often than users without an app (16% vs. 5%).

Table 3.6: Engaging with News Using a Mobile Twitter App.

	Have Twitter App	Do Not Have Twitter App
Follow journalist or news organization on twitter	50%	22%
Often click tweeted news	26%	9%
Often click trending news	25%	13%
Often retweet news	16%	5%

The number of days spent on Twitter also matters in engaging with news. Fifty-six percent of Twitter users who spend six to seven days a week on Twitter follow a journalist or news organization compared to the 21% of Twitter users who spend only one to two days per week. Thirty-seven percent of consumers who spend six to seven days per week on Twitter click tweeted news often but only 5% of consumers who spend one to two days on Twitter tweet news often (see Table 3.7).

Consumers who are on Twitter six to seven days per week are almost three times more likely than consumers on Twitter one to two days per week to click on trending news often. And although retweeting news is less popular than other news engagement on Twitter, consumers who are on Twitter at least six days a week are more likely to retweet often compared to consumers who spend one or two days on Twitter (23% vs. 3%).

While 2015 may go down in news media history as a watershed year for news outlets partnering with social media to increase news engagement, 2009 is remembered as the year Twitter's breaking news potential was first recognized.

Table 3.7: Twitter News Activity by Days on Twitter.

	1–2 Days Per Week	3–5 Days Per Week	6–7 Days Per Week
Follow journalist or news organization	21%	49%	56%
Often click tweeted news	5%	23%	37%
Often click trending news	11%	22%	31%
Often retweet news	3%	13%	23%

That was the year a passenger riding on a Hudson River ferry tweeted a photo of what became known as the "Miracle on the Hudson" (Krums, 2014).

> On January 15, 2009, US Airways Flight 1549, an Airbus A320 piloted by Captain Chesley B. "Sully" Sullenberger and First Officer Jeffrey Skiles, made an unpowered emergency water landing in the Hudson River after multiple bird strikes caused both jet engines to fail. All 155 occupants, the passengers and crew, successfully evacuated from the partially submerged airframe as it sank into the river; they were rescued by nearby watercraft. Several occupants suffered injuries, a few of them serious, but only one required hospitalization overnight. The incident came to be known as the "Miracle on the Hudson," and Captain Sullenberger was soon regarded as a hero by some accounts. (US Airways, n.d.)

The tweeted photo of the partially submerged plane with passengers and crew standing on the wings became part of Twitter lore and a defining moment for Twitter as a tool for breaking news. That tweeted photo only two years after Twitter won "best start-up in the blog category" at South by Southwest (SXSW) in Austin, Texas, showed how Twitter could be more than a status update (Bilton, 2013, p. 96). Today, Twitter is also a platform for breaking news, a medium for getting real-time news, and an indispensable reporting tool for professional and citizen journalists. Twitter has also evolved into a platform for ordinary people, celebrities, athletes, companies, politicians, voters, educators, non-profits, corporate leaders, organizations, and advocates to connect with others and attract media attention about an issue, idea, event, product, innovation, brand, civic concern, movement, or injustice. But it's the ability to create hashtags to facilitate connecting and finding tweets on a specific topic and tweet real-time messages, photos, and videos from the scene using a smartphone that has made Twitter into an indispensable tool for professional journalists and citizen journalists. And Twitter has become a platform for news consumers to find and engage with news in ways that were unimaginable only a half-decade ago. And it's this new Twitter-infused engagement with news that has had an effect on trust of a variety of sources of news.

Table 3.8 shows the Twitter effect by first comparing which has more of an impact on media trust: days per week on Twitter or percentage of time spent engaging with news on Twitter. For each and every news source—legacy, social media, interpersonal—spending at least one-third of Twitter time on news has more of an effect on trust than spending six to seven days per week engaging with news on Twitter.

Spending one-third of Twitter time engaging with news appears to affect trust of news sources in two ways: (1) increases level of trust for news source; (2) reorders sources trusted. The Twitter effect on level of trust is displayed in the last column as percentage point difference. Of eight news sources examined, Twitter benefits the most as a news source. Trust in Twitter as a news source increases from 9% to 42%, a 33 percentage point increase, when at least one-third of Twitter time is spent engaging with news. Twitter is not the only social networking site that benefits from Twitter time on news; Facebook also benefits. In fact, trust of Facebook for those who spend at least one-third of their Twitter time on news increases from 14% to 37%. With an increase of 20 percentage points in trust level (from 17% to 37%), journalists are the fourth-largest beneficiary of time consumers spend engaging with news on Twitter. And because of the importance of trust in the journalist–news consumer relationship, that 20 percentage point increase has significance beyond the actual number.

The Survey of Journalists in the Digital Age provides insight into why trust in journalists may have increased for consumers spending at least one-third of Twitter time on news. When journalists were queried about the impact of social media, 69% agreed or strongly agreed that social media helped them become more engaged with the audience and 30% agreed or strongly agreed that social media enhanced their credibility (Wilnatt & Weaver, 2014). Very likely this increased audience engagement and enhanced credibility via social media is translating into a higher level of trust for Twitter users spending at least one-third of their Twitter time engaging with news.

News sources that benefit the least from spending at least one-third of Twitter time on news are local TV News (+14 percentage points); network TV News (+12 percentage points); local online and print newspaper (+10 percentage points). The reason for a smaller increase in level of trust may be because local TV news, local online and print newspaper, and network TV news already had the highest trust levels when survey participants were asked how much they trusted each news source. Or it could be that by spending so much time engaging with news on Twitter, consumers felt that because Twitter allows independent, real-time access to news from multiple sources, local TV news, network TV news, and local online and print newspapers lose some of their advantage as a trustworthy news source.

Table 3.8: The Twitter Effect on News Source Trust.

	% Saying A lot	% on Twitter 6–7 Days Weekly	% Spending at Least 1/3 of Twitter Time on News	Impact of Spending at Least 1/3 of Twitter Time on News +/– Percentage Points
Trust Twitter as news source	9%	36%	42%	+33
Trust Facebook as news source	14%	24%	37%	+23
Trust national newspaper	26%	39%	47%	+21
Trust journalists	17%	28%	37%	+20
Trust family & friends as news source	28%	33%	43%	+15
Trust local TV news	42%	44%	56%	+14
Trust network TV news	30%	38%	42%	+12
Trust local print & online newspaper	34%	37%	44%	+10

As powerful as the Twitter effect may be on trust of news sources, Twitter's effect is not limited to trust. Twitter can also enhance enjoyment of news. For Twitter users who spend at least one-third of their Twitter time engaging with news, strongly enjoying news increases 20 percentage points, from 36% to 56%.

The importance of increasing trust and enjoyment should not be underestimated for the potential long-term effects on increasing news consumption. But news has to be available on social networking sites for it to have an effect. That's why news organizations have made their content available on every social media platform that has emerged in the news media landscape and that's why they are now partnering with Snapchat. But even if news is available, there's no guarantee users, especially the youngest Millennials and teens who have embraced Snapchat as their own, will even notice news to have an effect. In Chapter 10, we'll explore the potential of a Snapchat effect on news engagement by first determining whether news has been noticed.

· 4 ·

NEWS CONSUMERS' PREFERENCES IN A MOBILE ENVIRONMENT

In the 1990s, when the Internet first became a medium that consumers could access and navigate, newspapers began making their news available online by constructing news websites with home pages that functioned like a front page. An editor picked the most important news stories of the day for the homepage; less important news was relegated to other pages or never posted online. This journalistic tradition of deciding what was newsworthy and emphasizing the most newsworthy stories of the day, whether on the front page of a newspaper or homepage of a news website, makes editors powerful gatekeepers, a metaphor first used in a 1950 study to explain editors' decision making (White, 1950).

That 1950 case study of a newspaper editor's decisions about which news to select and which to reject also revealed the "subjective value-judgments" that editors used to determine what news the audience should see (White, 1950, p. 386). The all-powerful gatekeeper deciding what was news and the powerless reader reading what gatekeepers deemed important, defined the relationship between the editor and audience through the end of the 20th century. However, as the 21st century got underway, the editor-audience relationship began to change.

With the growing popularity of search engines and the emergence of Facebook in 2004 and Twitter in 2006, consumers began to discover that they didn't have to be captives of editors' "subjective value-judgments," they could get news on their own terms. No longer did consumers have to rely on editors' homepage picks. With the help of search engines and social media, consumers could choose news important to them and go find it directly on story pages. By 2012, when 75% of consumers did not first check in at a news website's homepage but went directly to a story page, Google's head of news and social products asked if news homepages still served a purpose (Perspectives, 2012, p. 13).

With an overwhelming majority of online consumers bypassing what was once the dominant destination for news online, it's logical to ask what purpose homepages serve today. It also makes sense to question the role and relevance of an editor during this power shift from editor to audience; that is, in today's news media landscape, the news the audience sees is no longer the exclusive domain of the gatekeeper's judgment.

The editor still has a role in the news that consumers are exposed to and engage with but that role is no longer as the only or even dominant gatekeeper. Today, gatekeeping is shared with search engines, social media, citizen journalists, tech companies, and even consumers themselves. This shared gatekeeping skips a news site's homepage and takes the consumer directly to a story page, diminishing an editor's gatekeeping authority. But when consumers use news apps, editors retain some of their power to determine what news consumers see because the entryway to news on apps is through the app's homepage, which lists the day's top stories. Still, apps are only one of many pathways to news on mobile.

In fact, when mobile device owners were asked about eight different pathways to news on mobile, 37% used an app to get news on their mobile device, thus seeing news prioritized by an editor (see Table 4.1). Still it must be emphasized that almost two-thirds of mobile news consumers likely did not see the editor's agenda of important stories of the day because they accessed news stories using browsers, search engines, social media, and notifications that connected to a story page—not a homepage reflecting the editor's judgment about that day's news.

Table 4.1: Pathways to News on Mobile Devices.

News apps	37%
Browser & search engine	27%
Social media	23%
Breaking news alerts & other notifications	11%
Other	1%

Benefits of News Apps

If platforms to access news were compared, the conclusion might be there is no quicker or more convenient way to get news than a mobile news app. You don't have to first turn on a TV or radio then search for news. There's no need to log on to your laptop or desktop. You don't have to first go to Facebook and let news find you. Or go to Twitter and scan news that's trending. Yet despite the quickness and convenience of apps, the majority of mobile news consumers do not use them. Even the design benefits are not enough for everyone to embrace and adopt apps. News apps are not just the mobile equivalent of news sites without clutter, many are elegant like a glossy magazine. Downloaded on a smartphone, apps are the entryway to news 24/7 with just a finger tap on the screen. With an app, news is scan-able and easily sharable on social media or by email or text.

News apps offer a mix of most important and recently updated stories; story topics are diverse, covering local, state, national, and world news. Apps for local TV news and newspapers cover cities, suburban communities, and even neighborhoods. And on an app like The BBC, the world is as close as a finger tap. An abundance of stories can be found for Europe, Latin America, the UK, Asia, the Middle East, and Africa.

News stories on mobile apps are almost always accompanied by a photo unlike stories on some mobile news sites that may only have a headline. Mobile apps also provide an index of subject categories to make it easy to find stories of interest. And no matter the subject, when news breaks, news apps usually announce the news with a breaking news banner at the top of the screen followed by updates as the news develops. News apps also have the capability of using sound and locked screen alerts for news that can't wait to be told.

The benefits of news apps no doubt contribute to the high grade given to news on mobile devices. When mobile news consumers were asked to grade the presentation of news on mobile, a whopping 70% gave a grade of "A" or "B." Mobile news consumers who used news apps were also more likely than mobile news consumers who did not use a news app to assign a grade of "A" to mobile news (33% vs. 26%; see Table 4.2). Furthermore, the use of a news app can reduce frustration that some may have trying to access news on a mobile device. When asked to agree or disagree with a statement about the frustration of reading news on a smartphone, news app users were more likely than nonusers of news apps to strongly *disagree* that reading news on a smartphone is frustrating (30% vs. 21%). News app use is also related to a perception that smartphones make it easy to keep up with news. According to Table 4.2, 52% of news app users compared to 39% of nonusers strongly agree that smartphones make it easy to keep up with news.

News apps also are related to enjoyment and regular engagement with news. More news app users (51%) than nonusers of news apps (37%) strongly agreed they enjoy keeping up with news and a larger percentage of news app users than nonusers of news apps get and seek news seven days per week.

Table 4.2: Impact of News Apps.

	Use News App to Access News	Don't Use News App to Access News
Strongly *disagree* reading news on a smartphone is frustrating	30%	21%
Strongly agree smartphone makes it easy to keep up with news	52%	39%
Strongly agree enjoy keeping up with news	51%	37%
Grade of A for news on smartphone	33%	26%
Get news 7 days per week	65%	55%
Seek news 7 day per week	47%	40%

As it relates to news, mobile-first consumers are also news app-first. Mobile-first consumers are more likely than mobile specialists and mobile laggards to use news apps. When news seeker specialists are separated from other mobile specialists, mobile-first consumers are still more likely to use news apps. Two-thirds (67%) of mobile-first consumers use news apps but only 46% of mobile news seeker specialists do. Mobile laggards are also news app laggards; only 39% use an app to get news.

In today's mobile world, there are not just more platforms for accessing news; on mobile devices, there are more ways than any other delivery platform to get news or be alerted to news. Plus, since news organizations first constructed news websites in the 1990s, there have never been more types of news conveniently available to consumers. Historically, legacy newspapers published seven broad categories of news: national; local, which included crime, courts, accidents, schools, city hall, and accidents; as well as international; weather; business; sports; and lifestyle news, once called women's pages, which included food, fashion, and features. National network news was heavy on government, politics, health, weather emergencies, and natural disasters. When CNN was launched in 1980, becoming the first 24-hour cable news network, its news offerings were primarily national. Sixteen years after the founding of CNN, two more 24-hour cable news networks were established: Fox News and MSNBC. But cable also offered specialized news: sports, weather, business, entertainment, and even legal. The sports cable news channel ESPN began in 1978, and over the next decade and a half, cable news channels devoted to weather, business, entertainment, and the law were launched. For every specialized news channel on cable, a specialized news website was created on the Internet once it became a place that consumers could access and navigate. Smartphones, tablets, and apps became the next frontier for established cable channels that offered specialized news and even specialized news websites that migrated from their Internet beginnings.

Whether general or specialized, published by legacy newspapers or broadcast by local TV, network TV, or radio, these genres of news are available in today's news landscape. But what is different today is that consumers have virtually unlimited access to the reporting of these topics. Using search, a web browser, social media, or a finger tap of a news app, anyone with a smartphone, wherever they live in the world, can access news on these news topics from CNN, BBC, the *Guardian*, the *New York Times*, *India Times*, *China Daily*, the *Los Angeles Times*, Fox, NBC, the *Washington Post*, NPR, AP, Huffington Post, or BuzzFeed to name a few sources.

While news genres have remained fairly constant over the years, how the news is reported and what various news organizations emphasize in these reports often varies. News may be targeted to specific age groups, generations, genders, races and ethnicities, income groups, professions, or political persuasions. Whether cable news, news websites, news available on social media or mobile, news targeted to attract specific groups may be framed to speak to that specific group. Framing to target news to specific groups may be more understandable if the framing is part of the news organization's mission and it's transparent about the framing. But when a news organization represents itself as upholding the values of journalism but then biases the news by conscious or unconscious framing, the public's confidence in journalism is undermined.

Framing, which scholars first applied to journalism in the early 1990s, can happen at the organizational level, such as Fox News, which reports news using a conservative frame (Tankard et al., 1991; Reese et al., 2001). Until MSNBC overhauled its approach to news in 2015 to emphasize breaking news, the NBC-owned cable channel reported news from a liberal or progressive perspective. Framing can also take place at a section or individual level where the section editor or reporter of an individual story frames the story so that it emphasizes certain aspects of the story or includes some people while excluding others or negatively stereotypes certain racial, ethnic, and religious groups.

The *Rolling Stone* rape story was not just an example of poor reporting; it was an example of extreme framing to emphasize a perspective about rape on college campuses. That retracted story not only hurt the credibility of the reporter and *Rolling Stone*, it hurt all of journalism by reinforcing a belief held by, perhaps, too many that journalists are biased.

And while the public may be skeptical of the press, the public is dependent on the press because it is the primary source for news. Whether getting news based on a gatekeeper's picks and framing posted on a news website or app homepage or going to a story page using social media, a browser, or notification, there is convenient access to more news published or broadcast by more news organizations than at any time in history.

But the convenient availability of more news from more sources does not mean that more types of news are getting news consumers' attention. Table 4.3, which lists the news consumers pay attention to most often, reminds us that despite the wide variety of news available today, only four types of news—local, weather, breaking, and national—get the attention of more than half of the public. Local, national, and weather news have always been popular and ubiquitous but historically breaking news was reserved for stories that were so big a network anchor would break into regularly scheduled programming to report news the public had to know immediately—for example, the 1963 assassination of President John F. Kennedy, the 1986 explosion of the space shuttle Challenger or the 9/11 terrorist attacks on the World Trade Center and Pentagon. And although in the pre-Internet era only TV and later cable news could break into regularly scheduled programming to report news, in the digital age there are no barriers to who can deliver news that is so important that it must be reported immediately.

Table 4.3: Most Popular Genres of News.

News Categories	Percentage Paying Attention Often
Local news	69%
Weather	66%
Breaking news	58%
National news	57%
International news	42%
Natural disasters	38%
Crime news	34%
Government news	32%
Politics	31%
Sports	30%
Election news	27%
Accidents	27%
Technology	23%
Health news	22%
Celebrity and entertainment news	21%
Environment news	20%
Consumer and personal finance news	19%
Business news	19%
Opinion about news	15%
Art and culture	13%

The Breaking News Effect

With news websites, mobile devices, and social media, legacy print news can break news immediately just as broadcast and cable have always been able to do. And with mobile devices consumers don't have to be in front of a TV; they can access breaking news wherever they may be. No wonder 58% of participants in the News for a Mobile-First Consumer Survey said they often pay attention to breaking news. News organizations' growing mobile-first approach to news, use of notifications to alert consumers to breaking news, and breaking news banners displayed on story pages no doubt have contributed to consumers' greater awareness of breaking news and increased attention paid to it. In fact, so much attention is paid to breaking news that it ranks number 3 on a list of 20 types of news behind local news and weather that have the attention of 69% and 66% of consumers, respectively.

Traditionally, breaking news has been reserved for unforeseen news about a threat, impending danger, major disaster, or death or troubles of a leader or icon. Fires, floods, tornadoes, hurricanes, earthquakes, plane crashes, mass shootings, terrorist attacks, and blizzards are often the subjects of breaking news that alerts the public about threats and danger. And while breaking news is usually a bad news alert, it should be noted that all breaking news is not bad, which was the case June 6, 2015, when history was made at the Belmont Stakes in New York. That's when American Pharoah raced into Triple Crown history, a sports achievement so rare that only 12 thoroughbred horses had accomplished it in 147 years (Triple Crown, n.d.). Because the last time there had been a Triple Crown winner was almost four decades before when Affirmed won in 1978, this horseracing achievement was not just news—it was breaking news. The *Washington Post*, *USA Today*, NBC News, CNN, and other news organizations posted breaking news banners on their mobile news app homepages almost immediately after American Pharoah crossed the finish line. The *News York Times* didn't have to post a breaking news banner because it reported the event live on its mobile app just like network TV.

Despite the ease with which consumers can get breaking news on smartphones and tablets, there's no statistically significant difference in getting breaking news between consumers who use a mobile device to engage with news and those who don't. And there's no difference in mobile consumer types. In other words, mobile laggards, mobile specialists, and mobile-first consumers are similar in getting breaking news. About three-fifths of each type of mobile consumer gets breaking news often. Surprisingly, there are no differences in attention paid to breaking news between Facebook users and non-users and Twitter users and non-users. Social media users and non-users pay about the same amount of attention to breaking news.

But there are factors that have an impact on attention to breaking news: seeking news and using a news app. And even some demographic characteristics are related to attention to breaking news. Daily news seekers are more likely than non-daily news seekers and people who never seek news to pay attention to breaking news often. Seventy-two percent of daily news seekers get breaking news often but only 50% of non-daily news seekers and 24% of never news seekers pay attention to breaking news often.

Mobile device owners who use a news app are more likely than mobile device owners who don't use a news app to pay more attention to breaking news; 68% of news app users compared to 56% of nonusers of news apps pay attention to breaking news often.

Some demographics do play a role in attention paid to breaking news (see Table 4.4). Baby Boomers are significantly more likely than younger generations to pay attention to breaking news often. Seven out of ten (68%) Baby Boomers compared to 59% of Generation X and 46% of Millennials often pay attention to breaking news.

Table 4.4: Summary of Demographics Related to Paying Attention to Breaking News Often.

Generation		
Baby Boomer	Generation X	Millennials
68%	59%	46%

Race and Ethnicity*				
White	African American	Asian American	Latino/ Hispanic	Native American
59%	64%	48%	54%	58%

Gender	
Male	Female
54%	61%

Income			
Under $40,000	$40–59,000	$60–79,000	$80,000+
55%	57%	58%	66%

Education*		
High School Degree or Less	Some College	College Degree Plus
55%	62%	57%

*Differences are not statistically significant.

Race and ethnicity may not distinguish level of attention paid to breaking news but breaking news still matters; the majority of each racial and ethnic group pays attention to breaking news often. Unlike race and ethnicity, gender distinguishes level of attention to breaking news. Furthermore, there is a surprising breaking news gender gap, with 61% of females compared to 54% of males paying attention to breaking news often. Income also matters in paying attention to breaking news; two-thirds (66%) of consumers making at least $80,000 annually pay attention to breaking news often, but only 55% of consumers making under $40,000 pay attention to breaking news often. Education level, like race and ethnicity, may not distinguish amount of attention paid to breaking news but breaking news is still popular.

Paying attention to breaking news can have negative consequences for how consumers evaluate the press. Consumers who paid attention to breaking news often (30%) were slightly more likely than those who paid attention to breaking news sometimes or less (25%) to strongly agree with the statement, "News organizations are more interested in being first than right." And because every news organization regardless of its legacy can break news today, there is competition to see who can be the first to break news, which can lead to mistakes that erode journalism's credibility. But breaking news also brings something new to get the attention of news consumers. And because breaking news transcends news category, news can break in almost all of the news categories listed in Table 4.3.

The list of news categories in Table 4.3 is not exhaustive, but it is still a reminder of the abundance of news offerings that consumers can pick from. For some, the list may be overwhelming because there are so many topics, but a factor analysis of the 20 news topics reveals essentially three non-overlapping clusters of news topics that consumers pay attention to. I have labeled these news clusters "local news watch," "watchdog news," and "personal benefit news" and listed the specific news topics under these clusters in Table 4.5. Local news watch includes the top three types of news paid attention to most by consumers: local, weather, breaking. Plus, it includes crime and accidents, two types of news heavily emphasized by local TV news.

Table 4.5: Three Clusters of News That Attract Consumers.

Local News Watch:	Watchdog News:	Personal Benefit News:
Local news	Government news	Art & culture
Crime news	Politics	Consumer news & personal finance
Breaking news	International news	Environment news
Accidents	National news	Celebrity & entertainment news
Weather	Election news	Opinions about the news
		Health news
		Technology news

Note: Factor Analysis produced three types of news that consumers pay attention to: Local News Watch, Watchdog News, and Personal Benefit News. Because the following news topics loaded high on at least two factors, they were excluded from the news clusters: business news, sports news, and disaster news.

One of journalism's most essential responsibilities is to be a watchdog of the powerful and the watchdog news cluster shows that consumers are paying attention to news that fulfils that journalistic responsibility because they are paying attention to national news, government, politics, elections, and so forth. Except for breaking news, the news in the local news watch cluster and watchdog cluster have been mainstays of journalism since journalism's earliest days. Even though there is no agreed-upon definition of hard news, most of the news in these two clusters is what is traditionally thought of as hard news as opposed to soft news (Poindexter & Harp, 2008). The third news cluster—personal benefit news—is evidence of a news consumer less interested in traditional hard news but interested in news, nonetheless.

There is no news website with a section labeled personal benefit news, yet most legacy news sites and digital native news sites post news included in the personal benefit cluster, and for some websites, a whole section is devoted to one of the topics. This personal benefit news is not what was once considered "soft" news in legacy newspapers. Kay Mills, author of *A Place in the News: From the Women's Pages to the Front Page*, used the "Four F's" to distinguish soft news from hard: "Hard news is news about foreign policy, the federal deficit, bank robberies. Historically, men's stuff. The right stuff. Soft news is news about the Four F's—family, food, fashion, and furnishings. Women's stuff. Back of the book. Plays, movies, books. Lifestyle" (Mills, 1988, p. 110).

As can be seen in Table 4.5, personal benefit news includes: art & culture, consumer news & personal finance, environment news, celebrity & entertainment news; opinions about the news; health news, and technology news. Today, health news is more personal health than policy, and technology is more about the social than technical, including social media, smartphones, and wearables. The opinions about the news category has expanded beyond newspaper editorials, op-ed columns, letters to the editor, and cable news talking heads. With comment sections provided by legacy and digital-only news media, as well as the opportunity for anyone to post an opinion about the news on Facebook, Twitter, and other social media platforms, opinions about the news have become an integral and indispensable component of the media landscape, and for some news organizations, comment sections are an "extension" of their "journalistic product" (Online commenting, n.d.).

Who is paying attention to these news clusters? To answer that question, the news topics that comprised the three news clusters were first summed then divided into three fairly equal parts, creating an index for each news cluster representing high attention, medium attention, and low attention. These news clusters were then analyzed by demographics. The highlights of that analysis are displayed in Table 4.6, which captures a departure from previous research and conventional wisdom about consumers' news preferences.

For decades there were no gender differences in news consumption but once a gender gap surfaced it tilted toward males over females (Poindexter, 2008a). While 46% of males compared to 30% of females frequently pay attention to watchdog news, the reverse is the case for the local news watch cluster; 31% of females compared to 25% of males often pay attention to the local news cluster. For personal benefit news, there is no gender difference.

Table 4.6: Demographics of News Clusters.

	Local News Watch	Watchdog News	Personal Benefit News
Gender	Female	Male	No difference
Generation	Baby Boomer Generation X	Baby Boomer	Generation X Millennials
Race	Black	White	Asian American
Education	High school or less	College plus	College plus
Income	No difference	$80,000 plus	$40,000 or less

Older generations pay more attention to the hard news clusters, with 46% of Baby Boomers paying attention to watchdog news compared to 35% of Generation X and 27% of Millennials. Baby Boomers (33%) and Generation X (30%) are also more likely than Millennials (21%) to pay attention to the local news cluster. However, Baby Boomers who are more interested in hard news are less likely to pay attention to personal benefit news compared to Generation X and Millennials. Only 27% of Baby Boomers pay attention often to personal benefit news, compared to 36% of Generation X and 32% of Millennials.

While race matters for all three news clusters, which race pays the most attention varies. Almost two-fifths (38%) of blacks pay high attention to the local news cluster; 40% of whites pay high attention to the watchdog news cluster, and 42% of Asian Americans pay high attention to the personal benefit news cluster.

The relationship with educational level varies by news cluster. For the watchdog and personal benefit news clusters, there is a positive relationship with those with at least a college degree paying more attention to this type of news than those with a high school degree or less. But for the local news watch cluster, adults with a high school degree or less pay more attention than the college educated. Unlike education, income level doesn't differentiate consumers in the local news watch cluster, but income level makes a difference in attention paid to watchdog and personal benefit news. Consumers with incomes of at least $80,000 pay high attention to watchdog news and consumers making $40,000 or less pay high attention to personal benefit news.

Demographics aren't the only factors differentiating attention paid to the three news clusters. According to Table 4.7, daily news seeking, mobile type, use of mobile in getting news, and social media are related to high attention paid to the local news watch, watchdog news, and personal benefit news clusters.

Table 4.7: News Seeking, Mobile, and Social Media Factors Related to News Clusters.

	Local News Watch Cluster	Watchdog News Cluster	Personal Benefit News Cluster
News seekers	Daily news seeker	Daily news seeker	Daily news seeker
Mobile types	All mobile types	All mobile types	Mobile specialists
Use mobile device for news	Both users and non-users of mobile device for news	Use mobile device for news	Use mobile device for news
Use news app	News app users	News app users	Both news app users and nonusers of news apps
Twitter user	Twitter users	Twitter users	Twitter users
Facebook user	Facebook users	Nonusers of Facebook	Facebook users
Instagram user	Both Instagram users and nonusers of Instagram	Both Instagram users and nonusers of Instagram	Instagram users

Not surprisingly, daily news seekers are more likely than non-daily news seekers to pay high attention to the local news, watchdog news, and personal benefit news clusters. Almost two-fifths (37%) of daily news seekers pay high attention to the local news cluster compared to 23% of consumers who seek news 1 to 6 days and 6% who never seek news. Personal benefit news has a similar pattern; 36% of daily news seekers compared to 5% of never news seekers pay high attention to personal benefit news. What most distinguishes watchdog news is that more than half (55%) of daily news seekers pay high attention to watchdog news compared to 26% of non-daily news seekers and 6% of consumers who never seek news.

There is no statistically significant difference among mobile-first, mobile specialists, and mobile laggards in attention paid to both the local news watch cluster and watchdog news, but differences do emerge for personal benefit news. Mobile specialists (38%) are more likely than mobile-first consumers (33%) and mobile laggards (26%) to pay high attention to personal benefit news often.

Using a mobile device to get news and using a news app make a difference in high attention paid to watchdog news. Consumers who use a mobile device for news (42%) compared to consumers who don't use a mobile device for news (35%) pay high attention to watchdog news. Using a news app also makes a difference; 51% of mobile device owners who use a news app pay attention to watchdog news compared to the 37% of mobile device owners who don't use a news app.

Use of a mobile device for news also matters in paying high attention to personal benefit news, but a news app does not. Two-fifths (41%) of mobile device owners who use their devices to get news pay high attention to personal benefit news but only 25% who don't use their mobile devices to get news pay high attention to personal benefit news. When it comes to the local news watch cluster, consumers who use their mobile devices for news and those who don't pay about the same high level of attention to this news cluster. But those who use a news app (37%) are more likely than non-users of news apps (25%) to pay a high level of attention to the local news watch cluster.

Table 4.7 also reminds us how integrated social media are in the news media landscape and that Twitter is a major player in paying attention to the three clusters of news. Twitter users are more likely than non-Twitter users to pay high attention to the local news watch cluster (31% vs. 26%), watchdog news cluster (41% vs. 36%), and personal benefit news cluster (40% vs. 25%).

While Facebook users are more likely than non-Facebook users to pay high attention to the local news watch cluster (30% vs. 23%) and personal benefit news cluster (32% vs. 25%), non-Facebook users are more likely than Facebook users to pay high attention to watchdog news (44% vs. 36%).

Concerning Instagram, there's no statistically significant difference between users and non-users in terms of paying high attention to the local news watch cluster and watchdog news, but there is a difference for personal benefit news. Instagram users (44%) are more likely than non-users of Instagram (26%) to pay high attention to personal benefit news.

In contrast to last century's legacy news landscape, today's 24/7 mobile terrain is rich, with a cornucopia of news platforms, assorted pathways to access news, and an array of news topics and approaches to tell news stories, including text, photo galleries, video, podcasts, infographics, animation, and virtual reality. In fact, with so much variety and access, one could say there's never been a better time to be a news consumer, especially because consumers are no longer dependent on gatekeeper editors for the news they have access to. With just a finger tap on a mobile screen, consumers can access news from anywhere in the world and on any topic. Perhaps what is best for news consumers is that by having access to any news, they also have access to the same topic from the perspective of different news organizations. With this access, consumers learn news organizations differ in how they frame stories, from the headlines to the sources and photos used. Consumers also have access to enterprise, breaking, and live streaming news from around the globe.

As consumers inform themselves about news produced by different news organizations, news organizations track consumer activity on their news sites, including which stories consumers click on or how much time they spend on a news site and where they go. These metrics, though, cannot measure attitudes and motivations that drive news engagement. Furthermore, because in today's world, it's virtually impossible *not* to get news even when you're not looking for it, knowing the percentage of people who get news is not as helpful as knowing the percentage who seek news and the reasons why. For example, in the News for a Mobile-First Consumer Survey, 57% of participants get news every day but that number drops 13 percentage points to 44% when the same consumers are asked about seeking news seven days a week. So to fully understand consumers in today's mobile news media landscape, including mobile-first, mobile specialists, and mobile laggards, it's necessary to dig deeper, which we will do in Chapters 5 through 8, where mobile, news engagement patterns and attitudes, and social media use are explored for Millennials, racial and ethnic groups, women, and groups from different political parties and persuasions.

· 5 ·

MILLENNIALS IN A MOBILE NEWS AND SOCIAL MEDIA WORLD

The dirty little secret about news consumption trends is that the youngest of the Millennial Generation are even less enamored of news than older Millennials, threatening the future of news engagement even more today than in 2012 when I first asked: Is news engagement a thing of the past? Declining news consumption, negative Millennial news coverage, Millennials' embrace of social media, and the changing of the generational guard have contributed to my concern about news engagement's future. The oldest generations that were committed to news are dying out and they are not being replaced by the Millennial Generation. Legacy news coverage about Millennials may be less negative today than four years ago but it still routinely excludes them from the news, reinforcing the message that legacy news and even many non-legacy news sites don't care that much about them even though they are critical to their future.

During its short history, social media's relationship with Millennials has been the opposite of legacy media's; from the beginning social media embraced Millennials and as more social networking sites emerged, social media attracted more of Millennials' attention and filled up more of their time, time that could have been spent with news. And just as social media appeared to reach a plateau in attracting older Millennials' attention as they came of age, smartphones emerged and commanded the attention of younger Millennials, giving them round-the-clock access to all things not news, further diminishing the potential that news could play in their daily lives.

As the fifth generation to be born in the 20th century, Millennials, who were born between 1983 and 1999, are the most different from the four generations that came before them: Greatest Generation (born 1901–1924); Silent Generation (born 1925–1945); Baby Boomers (born 1946–1964); Generation X (born 1965–1982) (Poindexter, 2012). Every generation is defined by the years they were born as well as the historical and cultural markers that leave a lifelong imprint on their lives, that is, if the individual is old enough to be affected. For example, the youngest Millennials were toddlers when on September 11, 2001, terrorists commandeered commercial jets and flew them into the Twin Towers of the World Trade Center in New York City and the Pentagon in Washington, D.C., while the oldest Millennials were graduating high school and starting college. As the youngest Millennials grew up, they learned about 9/11 from parents, teachers, history books, or a trip to the 9/11 Memorial Museum in New York City, while the oldest Millennials who were coming into adulthood experienced 9/11 by watching the shocking breaking news live on TV or viewing replays of the crumbling Twin Towers during local, cable, and network newscasts. While the youngest Millennials do not have firsthand knowledge that would shape a personal memory, the oldest Millennials do because they, like older generations, know what 9/11 did to American society:

> September 11, 2001, touched every aspect of American society. It destroyed lives and devastated families. It ravaged property, wiped out businesses and crippled the American economy. September 11 changed the President's agenda and re-ordered the priorities of Congress, started a war and established new laws and regulations that tightened airport security and immigration. (Poindexter & Conway, 2005, p. 175)

To illustrate, the youngest Millennials have only known post-9/11 airport security. The youngest Millennials have never experienced being dropped off or met at an airport gate by a friend or loved one. The youngest Millennials, unlike the oldest Millennials, have always had to show a photo ID, take off their shoes, pull out their laptops, and go through the body-scanning equipment before entering the airport's secure area. This difference between the 9/11 experiences of the youngest and oldest Millennials underscores the importance of recognizing differences within generations:

> Usually when generations are reported on or talked about, the descriptions are referring to a group born during years spanning a decade and a half to almost two decades. These reports rarely emphasize that within a generational birth period there are two distinct groups that I call Wave I and Wave II. This distinction is important because a closer look at these two sub-generational groups reveals differences that can both distinguish within a generational birth period and overlap a portion of a generation that preceded or followed it. (Poindexter, 2012, p. 23)

In addition to historical and cultural differences across and within generations, there are demographic, attitude, and value differences that distinguish Millennials from previous generations. The two demographics that most distinguish Millennials are race/ethnicity and education; Millennials are the most racially and ethnically diverse generation and have more education than generations that preceded them (Poindexter, 2012). Additionally, as a generation, Millennials have different attitudes and values than their forebears; they are more interested in "making a difference than a dollar" (Rampell, 2011). Generation experts Neil Howe and William Strauss identified seven characteristics that distinguished the "Millennial persona": "special, sheltered, confident, team oriented, achieving, pressured conventional" (Howe & Strauss, 2000, pp. 43–44). One of the first Pew Research Center reports on Millennials described them as more likely to have multiple tattoos, to be more tolerant, and to be more receptive to immigrants (Millennials, 2010).

Within the Millennial Generation, the Great Recession of 2008 had more of an impact on Wave I Millennials than Wave II Millennials. Just as Wave I Millennials were graduating college, the Great Recession hit, reducing job opportunities as well as the chance for them to live on their own. Rather than being able to rent an apartment, many Millennials had to move home after college graduation. Student loan debt also contributed to the economic woes of Wave I Millennials; being unemployed or underemployed made it harder to re-pay student loans (Marks & Martin, n.d.).

But there were some generational benefits Wave I Millennials were the first to experience. After passage of Obamacare, Wave I Millennials were the first to be carried on their parents' insurance until age 26, one year beyond what the law previously allowed (Young adults and the Affordable Care Act, n.d.). Millennials were also the first to have their student loan repayment structured according to their income, and they were the first generation to have their student loan debt forgiven if they worked in the public sector for 10 years (How to repay your loans, n.d; Public service loan forgiveness, n.d.).

Now 83.1 million strong, Millennials are the largest living generation, and according to the Pew Research Center, since March 2015, there are more Millennials in the workplace than Generation Xers or Baby Boomers (White, 2015; Fry, 2015). And in college classrooms, younger Wave II Millennials have replaced older Wave I Millennials, who have started turning 30 and have been in the job market at least one decade.

Millennials—like every generation born in the 20th century—has a name assigned by a marketer, the media, or a generation expert, but it's rare for generations to be called two different names that compete in the media and lead to confusion and a lack of appreciation for a generation's uniqueness and potential. Fortunately, the label Millennials appears to have won out over the non-distinctive Generation Y label that was more about following Generation X than recognizing both the unique attributes of this generation and the period in which it was born. The Millennial label, which was coined by generation experts Neil Howe and William Strauss (2000) in *Millennials Rising: The Next Great Generation*, paid homage to the new 21st-century and new millennium in which this generation would come of age. And knowing what we now know about this generation that was born in the 20th century but came of age in the 21st century, Millennials are so different from Generation X that it is inaccurate to saddle them with a label that suggests they are following in the footsteps of Generation X, which couldn't be further from the truth.

The media and technology available to and adopted by Millennials have also set Millennials apart from older generations and within the Millennial generation. In the case of personal and Millennial-targeted media and technology, perhaps, nothing has distinguished Millennials more from older generations. As can be seen from Table 5.1, during the formative years of each generation born during the 20th century, a media or technological innovation emerged that had an impact on them and transformed that generation and the generations that followed.

Of course, some innovations like TV and smartphones, because of their rapid adoption, had more of an impact than others. And while the attributes of an innovation played a significant role in adoption, the life stage—kid, teen, young adult—also mattered for the adoption of some innovations. The telephone, radio, movies, and even the first mass-marketed camera had a far-reaching effect on the Greatest Generation, born 1901–1924. TV transformed the Baby Boomer Generation and while the personal computer emerged during the early years of Generation X, it wasn't until it became kid and classroom friendly after the Millennial Generation was born that it really took off. The World Wide Web emerged as the Millennial Generation was growing into its teenage years but it was through the family-friendly America Online that Millennial kids first ventured online (Poindexter, 2012). While Facebook wasn't the first social networking site, by launching as the first wave of Millennials went off to college, Facebook was rapidly adopted as it rolled out across college campuses, forever linked to the Web and computer savvy Millennials as they were coming of age. And even though two years after Facebook was founded, anyone 13 or older could join Facebook, the oldest Millennials continue to be identified with that social network (Kirkpatrick, 2010).

Table 5.1: Kid, Teen, and Coming-of-Age Media and Technology Innovations Adopted by Five Generations.

Wave II Millennials Born 1992–1999	Smartphones, tablets, apps, mobile social media, mobile news, mobile banking, online shopping, streaming music, streaming video, diverse emoji, cloud storage, wearables
Wave I Millennials Born 1983–1991	Handheld games, the Internet, World Wide Web, music downloading, cellphones, texting, YouTube, news websites, social media, search engines, Wikipedia, citizen journalism, blogs, *The Daily Show with Jon Stewart*, laptops, IM, iPod, Netflix, Redbox, camera phones, selfies, streaming movies, entertainment/celebrity news, online shopping, online banking, debit cards, online gaming, lol, omg, video chat (Skype, FaceTime) digital camera, laptop, news website, online dating
Generation X Born 1965–1982	MTV, 24-hour cable news, personal computers, VCRs, renting movies, reality TV, ATMs, camcorders, video games, *Sesame Street*
Baby Boomers Born 1946–1964	TV, kid- and teen-targeted programming, including variety and dance shows and cartoons
The Silent Generation Born 1925–1945	Polaroid camera, comic books, *Seventeen*
Greatest Generation Born 1901–1924	Radio, newspaper extra, movies, newsreels, telephone, Brownie camera

As younger Wave II Millennials grew into their teen years, there was less of an identification with Facebook, perhaps because their parents and grandparents had joined the social networking site once exclusive to Wave I Millennials, or maybe it was because at the same time that younger Millennials grew into their teenage years, they experienced the coolest technology ever: the iPhone. Older Millennials had come of age on cellphones but that mobile phone was more about texting and phone calls. The iPhone, though, had a "wow factor," and as Wave II Millennials came into their teen years, their parents bought them an iPhone, and as Wave I Millennials graduated college and started their first real jobs, cellphones were replaced with smartphones that had many more features than texting and phone calling.

Plus, the iPhone was easy to use, elegant in its design, and fun. And because a variety of generation-changing technologies (camera, camcorder, radio) and access to movies, TV, music, games, social media, search, and information had been bundled into one device that fit into one's pocket or purse, the smartphone became indispensable, which is why it transformed the Millennial Generation in a way that no previous generation had been transformed. For Wave II Millennials, a smartphone was more than a mobile device; a smartphone became part of Wave II Millennial's identity just like Facebook became part of Wave I's identity. In fact, smartphones became so much of an extension of who Wave II Millennials were that many didn't even bother to carry them in their pockets, purses, or backpacks; Wave II Millennials just carried their smartphones in their hands, connected to earbuds and headphones listening to streaming music, talking on the phone, sending and receiving texts, checking social media and YouTube, searching, sharing, checking directions and prices, checking in at the airport, or buying a latte at Starbucks. And as an extension of their identities, Millennials personalized their smartphones with the cases they purchased and the wallpaper they chose for the lock and home screens. But personalization didn't stop there. Millennials also personalized ringtones and the sounds from alerts, new mail, texts, tweets, and push notifications.

There's no question that Millennials and mobile have become one—but which mobile identity—mobile-first, mobile specialist, mobile laggard—best describes Millennials and is the identity the same for older and younger Millennials? Currently mobile-first consumers like innovators in the adoption model represent a small percentage of consumers, but even of this small percentage, Millennials are more likely to be mobile-first (see Table 5.2). Mobile-first is not just about the smartphone; it's also about behavior and attitude.

Compared to other technology, how important are smart mobile phones in the lives of Millennials? A whopping 96% of the youngest Millennials, according to Bank of America's Trends in Mobility Report, cited their mobile phone "as most important to their daily lives" (Trends, 2014). Other technologies, although important, were not as important as the smartphone: Internet (88%), personal computer (82%), social networking site (56%), and TV (54%).

Millennials are also far more likely than older generations to be mobile specialists. Over half (53%) of Wave II Millennials and two-fifths (40%) of Wave I Millennials are mobile specialists compared to about one-third (34%) of Generation X and one-fifth (22%) of Baby Boomers. And not surprisingly, Wave II Millennials are less likely to be mobile laggards; 44% of Wave II Millennials are mobile laggards compared to 77% of Baby Boomers. With the overwhelming importance of smartphones to the youngest Millennials, it won't be long before a larger percentage are mobile-first, further transforming their behavior and attitudes, which is why institutions, organizations, companies, and legacy news media must get out in front of this mobile-first tsunami that is on the horizon. But will they?

Table 5.2: Generations and Mobile Consumer Types.

	Wave II Millennials (Born 1992–1999)	Wave I Millennials (Born 1983–1991)	Generation X (Born 1965–1982)	Baby Boomers (Born 1946–1964)
Mobile-first	3%	5%	2%	1%
Mobile specialists	53%	40%	34%	22%
Mobile laggards	44%	55%	64%	77%

Just as laggards exist among mobile consumers, laggards exist among institutions, organizations, companies, and legacy news media. Over the course of the digital revolution that disrupted how news was reported and consumed and how money was made, legacy news organizations were slow to respond and, in some cases, resistant to the inevitable digital overhaul of the news landscape. And that laggard pattern may be repeating itself despite consumers' rapid adoption of mobile. As the largest living generation leads the mobile-first revolution, news organizations cannot afford to be laggards or continue a tradition of negative news coverage of Millennials, that is, unless news organizations want fewer—not more—news consumers in the future.

Historically, demographic, socialization, motivation, avoidance, belief, and attitude factors have played roles in whether consumers engage with news or not (Poindexter, 2008c). In the case of the Millennial Generation, social media and negative news coverage of their generation have also been influential in their lack of news engagement (Poindexter, 2012) but in strikingly different ways. Social media embraced Millennials and Millennials embraced social media back, leaving less time for news, even for those who might have been interested in news. Legacy news, on the other hand, didn't reach out to Millennials and didn't embrace them. In fact, legacy news media repelled Millennials with negative news coverage about their generation. The news media stereotyped Millennials, excluded them, and sometimes insulted them. That's why when Millennials were asked to grade the news media's coverage of their generation, 68% gave a grade of C, D, or F (Poindexter, 2012).

The news coverage about Millennials that most deserved a failing grade was the infamous 2007 *60 Minutes* broadcast, "The Millennials Are Coming" (2007). Morley Safer, who reported the segment, called Millennials a "new breed of American worker" who attacks everything held sacred. Labeled "narcissistic praise-hounds," the *60 Minutes* segment said the problem with Millennials was that children's TV host Mr. Rogers called them "special when they were preschoolers." The *60 Minutes* broadcast also said Millennials had not been trained "how to eat with a knife and fork" and had not been trained "how to work in an office because they've never held a summer job." Finally, the broadcast criticized Millennials for "moving home after graduation."

Wall Street Journal Pulitzer Prize–winning columnist Bret Stephens may have outdone *60 Minutes* in his column to the Millennial graduates of 2012:

> Most of you have spent the last few years getting inflated grades in useless subjects in order to obtain a debased degree. Now you're entering a lousy economy, courtesy of the very president whom you, as freshmen, voted for with such enthusiasm. Please spare us the self-pity about how tough it is to look for a job while living with your parents. They're the ones who spent a fortune on your education only to get you back— return-to-sender, forwarding address unknown. (Stephens, 2012)

The *Wall Street Journal* columnist wasn't finished with insulting Millennials. In addition to calling them the "least knowledgeable graduating class in history," he criticized their résumés by adding: "Your prospective employers can smell BS from miles away. And most of you don't even know how badly you stink" (Stephens, 2012).

What most set the *60 Minutes*' Millennials segment and the *Wall Street Journal* column apart from other negative news about Millennials was that they were mean-spirited in their insults of the generation born in the 1980s and 1990s. Most of the news coverage of this generation has been negative—not mean-spirited—but negative news does not endear itself to Millennials. In the book *Millennials, News, and Social Media* (Poindexter, 2012), Millennials complained that their generation was routinely excluded from the news unless they were a celebrity, had committed a crime, were in an accident, or the story focus was on graduations. News purportedly about Millennials was also criticized for not including quotes from Millennials or even photographs (Poindexter, 2012). Plus, Millennials criticized the news media for failing to report stories of interest to Millennials.

Since the mean-spirited 2007 *60 Minutes* segment and 2012 *Wall Street Journal* column, Millennial news coverage has been less overtly negative, new news platforms targeting the Millennial Generation have become part of the news landscape, and a *New York Times* columnist admitted that older generations that preceded Millennials had, indeed, bashed them:

> Dear Millennials, We're Sorry
>
> Among Americans age 40 and older, there's a pastime more popular than football, Candy Crush or HBO. It's bashing millennials. Oh, the hours of fun we have, marveling at their self-fascination and gaping at their sense of entitlement! It's been an especially spirited romp lately, as a new batch of them graduate from college and gambol toward our cubicles, prompting us to wonder afresh about the havoc they'll wreak on our world. We have a hell of a lot of nerve, considering the havoc we've wrought on theirs. (Bruni, 2014)

The embrace by social media and bashing by legacy news media have taken a toll on the relationship between Millennials and news. Whether measured as "getting" or "seeking" news, Wave I and Wave II Millennials are less likely than older generations to be daily news consumers. While one-third of the youngest Wave II Millennials get news daily compared to three-fourths of Baby Boomers, only 17% of Wave II Millennials *seek* news daily compared to 60% of Baby Boomers. Table 5.3 emphatically underscores the decline of news consumption across three generations born in the 20th century and raises concerns about the future of news engagement.

Millennials may be less likely than older generations to get or seek news every day, but Table 5.3 does offer some optimism because Millennials do pay attention to some types of news often. The youngest Wave II Millennials pay more attention than Wave I and older generations to two types of news, according to Table 5.3. More than one-third (36%) of Wave II Millennials

Table 5.3: Millennials, Older Generations, and News Consumption and Attitudes.

	Millennials Wave II	Millennials Wave I	Generation X	Baby Boomers
Daily News Activity:				
Get news	33%	40%	50%	75%
Seek news	17%	28%	37%	60%
Often Pay Attention to:				
Breaking news	50%	44%	58%	68%
Local news	49%	56%	71%	78%
National news	45%	41%	55%	68%
Celebrity & entertainment news	36%	28%	22%	15%
Technology news	32%	25%	25%	18%
International news	29%	33%	41%	47%
Government news	24%	26%	30%	39%
Politics	21%	22%	29%	38%
Election news	18%	19%	26%	34%
Strongly Agree Attitudes:				
Duty to keep informed	35%	29%	36%	40%
Raised to believe being informed is important	25%	28%	32%	41%
People important to you value keeping up with news*	14%	24%	22%	18%
Enjoy keeping up with news	26%	30%	34%	40%
Depend on news for daily life	9%	18%	17%	16%

*Differences are not statistically significant.

often pay attention to news about celebrities and entertainment while only 15% of Baby Boomers pay attention to this genre of news. And almost one-third (32%) of Wave II Millennials pay attention to technology news compared to 18% of Baby Boomers. But the news genres that get most of Wave II Millennials' attention are breaking (50%), local (49%), and national (45%). Local news is of most interest to older Wave I Millennials, with 56% getting this type of news often. Breaking news is paid attention to most often by 44% of Wave I Millennials and 41% often pay attention to national news.

Of the nine types of news examined in Table 5.3, news about government, politics, and elections hold the least interest for younger and older Millennials. About a quarter of Wave II and Wave I Millennials often pay attention to news about government compared to 39% of Baby Boomers. Only 21% of Wave II Millennials and 22% of Wave I Millennials often pay attention to news about politics compared to 29% of Generation X and 38% of Baby Boomers. Less than one-fifth of Wave II and Wave I Millennials often pay attention to election news compared to one-quarter of Generation X and one-third of Baby Boomers. Government, politics, and election news are usually what's meant by hard news and they are generally less relevant to Millennials who are in the first stage of adulthood. This type of news also has the potential to be the most boring because reporters who cover this type of news are often writing for knowledgeable readers, experts, officials, politicians, and other journalists. But this news is some of the most important because hard news allows journalists to perform their watchdog function.

Millennials not only pay less attention than older generations to government, political, and election news, they also are more likely than any generation to get news about politics and government from social media—not legacy media. A Pew Research Center study found Facebook is the place 61% of Millennials get news about politics and government while local TV is where 60% of Baby Boomers go for this type of news (Mitchell et al., 2015).

Because Millennials often pay attention to some types of news, they are aware of what's in the news, including whether or not they are covered and how they are treated. When news organizations exclude or disparage Millennials, it discourages Millennial engagement with news and devalues the importance of news in their lives. To assess the value placed on news and being informed, participants in the News for a Mobile-First Consumer Survey were asked how much they agreed or disagreed with five statements. As can be seen in Table 5.3, none of the five statements about news and being informed was strongly endorsed by the majority of any generation but there were significant differences among the generations. Wave I Millennials (29%) were least likely to strongly agree they had a duty to be informed, while Baby Boomers (40%) were most likely to strongly agree that being informed was a duty. Only one-quarter (25%) of Wave II Millennials strongly agreed they were "raised to believe being informed is important," while 41% of Baby Boomers strongly agreed with that statement. Ironically, the youngest Wave II Millennials and oldest Baby Boomers do not have people who are important to them that value keeping up with news. Only 14% of Wave II Millennials and 18% of Baby Boomers strongly agreed with this statement.

Perhaps the most telling statements about how Millennials feel about news are whether or not they enjoy news and whether they depend on news in their daily lives. Wave II Millennials are least likely to enjoy keeping up with news and the least likely to depend on news in their daily lives. Slightly more than one-fourth (26%) of the youngest Wave II Millennials enjoy keeping up with news compared with 40% of Baby Boomers. While only 9% of younger Wave II Millennials depend on news in their daily lives, news does not appear to be that essential in the daily lives of Wave I Millennials and other older generations; less than one-fifth of Wave I Millennials (18%), Generation X (17%), and Baby Boomers (16%) depend on news in their daily lives.

Table 5.4 reminds us that the relationship between Millennials and social media is very different from the relationship between Millennials and news and so are the patterns between social media and the younger and older generations. Plus, it underscores that Wave I and Wave II Millennials are not the same when it comes to social media.

At least four-fifths of younger and older Millennials as well as Generation X are on Facebook compared to the two-thirds (67%) of Baby Boomers. Wave II Millennials are significantly more likely to be on the photo-sharing Instagram social networking site; almost two-thirds (65%) of Wave II Millennials compared to just under half (49%) of Wave I Millennials, 29% of Generation X, and 5% of Baby Boomers are on Instagram.

Wave II Millennials are also almost three times more likely than Baby Boomers (58% vs. 21%) to be on Twitter. Additionally, younger Millennials (58%) are more likely than older Millennials (48%) and Generation X (43%) to be on Twitter.

While Wave II Millennials are more likely than older generations to be on Twitter, they are less likely than Wave I Millennials and Generation X to spend their time on Twitter engaging with journalists or news. More than half

(51%) of Wave I Millennials and 48% of Generation X follow a journalist on Twitter compared to 38% of Wave II Millennials and 24% of Baby Boomers. Wave II Millennials (19%) are also less likely than Wave I Millennials (28%) to click on tweeted news.

Both younger Wave II Millennials and older Wave I Millennials depend on social media in their lives more than Generation X and Baby Boomers. Younger Wave II Millennials depend on social media twice as much as news (18% vs. 9%), while older Wave I Millennials depend on social media and news about the same (19% vs. 18%).

Table 5.4: Millennials, Older Generations, Social Media, and Mobile News.

	Millennials Wave II	Millennials Wave I	Generation X	Baby Boomers
Social Media Activity:				
On Facebook	86%	83%	81%	67%
On Instagram	65%	49%	29%	5%
On Twitter	58%	48%	43%	21%
Follow journalist or news organization on Twitter	38%	51%	48%	24%
Often click tweeted news	19%	28%	24%	12%
Depend on social media for daily life	18%	19%	13%	7%
Mobile News Activity and Attitudes:				
Use mobile device to engage with news	67%	62%	54%	27%
Strongly agree mobile devices make it easy to keep up with news	37%	34%	33%	20%
Use app to engage with news*	28%	37%	40%	37%

*Differences are not statistically significant.

Just as there are generational differences in terms of news and social media, there are also generational differences with mobile news, as can also be seen in Table 5.4. Younger Wave I (62%) and Wave II Millennials (67%) are more likely than Generation X (54%) and Baby Boomers (27%) to use mobile devices to engage with news. Wave II Millennials are also more likely than older Wave I Millennials, Generation X, and Baby Boomers to strongly agree mobile devices make it easy to keep up with news. But there is no significant difference among younger and older generations in using an app to engage with news.

While all generations give high marks to mobile news, a larger percentage of Wave I and Wave II Millennials than older generations give grades of A or B to mobile news (see Table 5.5). Over three-quarters (76%) of Wave I and Wave II Millennials give a grade of A or B to mobile news; 70% of Generation X and 62% of Baby Boomers assign an A or B grade to mobile news.

To put the high mobile news grade in perspective, Table 5.5 compares that grade with grades for news coverage in general and news coverage of the most important problems. Of the three types of news, mobile news is graded the highest and the youngest generation gives higher grades than older generations. Between 48% and 54% of each generation gives grades of A or B to news coverage overall but grades for news coverage of most important problems are less generous. Still slightly less than half (48%) of Wave II Millennials grade overall news coverage of most important problems an A or B but the percentage of Baby Boomers giving a grade of A or B to this type of news drops to 36%.

Table 5.5: Grading Mobile News, Overall News Coverage, and News of Most Important Problems.

	Millennials Wave II	Millennials Wave I	Generation X	Baby Boomers
A/B grades:				
Mobile news	76%	76%	70%	62%
News coverage overall	52%	54%	49%	48%
News coverage of most important problems	48%	42%	41%	36%
D/F grades:				
Mobile news	4%	6%	4%	6%
News coverage overall	9%	16%	19%	20%
News coverage of most important problems	15%	29%	29%	31%

The failing and almost failing D and F grades may be more telling than the high A and B grades because bad grades point out problem news coverage in ways that most people can relate to. While few failing grades were given for mobile news, for each generation, an increasing percentage of survey participants gave D and F grades to overall news coverage and news coverage of most important problems. In fact, for both younger and older generations, four to seven times as many D and F grades were given to news coverage of most important problems as were given to mobile news.

As news competes with social media for the attention of Millennials and fewer Millennials value news and the importance of being informed, mobile devices may be what's keeping the Millennial-news relationship alive. This now tenuous relationship between news and news consumers hasn't always been this way.

Long before online news, social media, and mobile devices emerged in the news media landscape, newspapers, TV, radio, and magazines were the only places to go for news. During this era when legacy media reigned supreme, a seminal *Journalism Quarterly* study declared newspaper reading a "thoroughly institutionalized behavior of Americans" (Westley & Severin, 1964, p. 45). It was 1964 and 86% of Americans read a newspaper daily. The year the article was published the first Baby Boomers started turning 18, coming into the first stage of adulthood. It was also one year after the assassination of President John F. Kennedy in Dallas, Texas, and the swearing-in of President Lyndon Johnson. The year 1964 also signaled the beginning of a new era following a decade of non-violent Civil Rights protests marred by the South's violent resistance to integration and the right of blacks to vote. That year Congress passed and President Johnson signed the Civil Rights Act that outlawed segregation in public facilities. One year later, the Voting Rights Act became the law of the land, making it illegal to keep blacks from voting. But the promise of the dream that Dr. Martin Luther King, Jr. articulated in his "I Have a Dream" speech appeared to some to be just a dream when on April 4, 1968, he was assassinated in Memphis, Tennessee.

All of these events were carried on the front pages of the Black Press and mainstream newspapers. These events were the cover stories of newsweekly magazines and the top news on radio and TV, which no doubt contributed to the fact that as Baby Boomers were growing into adulthood, almost nine out of ten Americans read a newspaper every day. Today, there are more ways to get news than at any time in history and there is more news. So why is it that a half century after daily newspaper reading was called a thoroughly institutionalized behavior only one-third of the youngest Millennials get news daily and fewer than one-fifth regularly seek news?

This half-century news consumption decline across two generations cannot be attributed solely to attention devoted to social media and smart mobile devices because these innovations did not emerge until after the oldest Millennials became adults. Some of the decline is likely due to the falling-off in the socialization of younger generations about the importance of being informed (Poindexter & McCombs, 2001).

Since socialization is the "social process by which individuals come to belong to a society and acquire some of its values, beliefs, perspectives, knowledge, social norms, and preferences—in short, its culture" (Wright, 1986, p. 185), the belief that being informed is important must be passed on and acquired. But that socialization appears to be waning, according to Table 5.3. While 41% of Baby Boomers strongly agreed with the statement that they were "raised to believe being informed is important," only 25% of Wave II Millennials were raised to believe it is important to be informed.

Parents, though, are not the only ones responsible for socializing the next generation about the importance of being informed. In a democratic society, educational institutions, from pre-school to higher education, as well as civic and political bodies, have a responsibility to develop informed citizens. And, perhaps, no individual or entity has more responsibility than journalists and news organizations because their business is "to provide people with the information they need to be free and self-governing" (Kovach & Rosenstiel, 2014, p. 9). And journalism educators have a responsibility because they are preparing the next generation of journalists who will report the news the public requires in a democracy.

Sharing responsibility to reinvigorate the belief that it is important to be informed is only part of the challenge. There must also be news worthy of being informed about, and there must be an effort to reach out to younger generations, especially during life stages that matter the most in creating future news consumers: kids, teens, and young adults. Parents and teachers have the most influence on kids but by the teenage and young adult years, peers and media are often more influential. That's why it should be of concern that only 14% of the youngest Wave II Millennials say people who are important to them value being informed. In other words, a whopping 86% of the youngest Millennials say the people who are important to them do not value keeping up with news.

Legacy media could have played a role in changing this statistic but that's not what legacy media did. While the authors of *The Elements of Journalism* say legacy media "cannot take all the blame," the authors admit the news business had a "business strategy that helped create non-news consumers" (Kovach & Rosenstiel, 2007, p. 211). And because Millennials have come of age during the 21st century, with their time and attention devoted to social media and smart mobile devices, the waning commitment to being informed and declining news consumption will be difficult to reverse.

The Millennial News Landscape

Even if legacy media did little to engage Millennials and even sometimes repelled them with negative news coverage about their generation, others, including digital-only media such as BuzzFeed, national legacy media businesses such as the *New York Times*, journalism professors, and Millennials have been trying to reverse this trend of a weakened commitment to being informed and declining news engagement with news by creating news platforms that appeal to Millennials. Only time will tell whether these Millennial-targeted news innovations, which include BuzzFeed, HuffPost College, "Millennials and News," NYT Now, Snapchat's Discover, Social News Network, theSkimm, and USA TODAY College, are too little too late (see Table 5.6). It should be noted that although neither former *New York Times* executive editor Jill Abramson nor NYT Now editor Cliff Levy described NYT Now as a Millennial news product, I know the app's features appeal to Millennials, which is why I have included it in Table 5.6 with other news products created to engage Millennials (AEJMC, 2014b; Sebastian, 2015a).

Table 5.6: News Platforms That Engage Millennials.

News	Year Founded	Mission/Target	Platform	Frequency of Updating
BuzzFeed	2006	Targets under 40 with listicles, sharable news, and videos that go viral	Site and mobile app	Daily with updates
HuffPost College	Huffington Post founded 2005	College students	Blog on Huffington Post	Varies
"Millennials and News"	2011	News curated by a journalism professor and Millennial that Millennials should know; news about and of interest to Millennials	Facebook mobile app http://on.fb.me/Nlejee	Daily and occasional breaking news

Table 5.6: (*continued*)

News	Year Founded	Mission/Target	Platform	Frequency of Updating
NYT Now from the *New York Times*	Founded as paid app in 2014; converted to free in 2015	Curates news from the *New York Times* and other sites for mobile consumers who are much younger than the general *New York Times* audience	Mobile app	Daily with daily updates
Snapchat's Discover	2015	News produced by media for Millennials	Snapchat mobile app	Varies
Social News Network	2012	Social media-driven publication founded by a journalism professor; curated by college students for college students	Twitter, Facebook, Pinterest, Snapchat, and Tumblr	Daily except semester breaks
theSkimm	2012	News curated by Millennials with a Millennial sensibility	Email	Monday–Friday Mornings
USA TODAY College	2011	News of interest to college students produced by college students	Site College. usatoday. com	Varies

BuzzFeed Brings a Non-Traditional Approach to News

BuzzFeed, which was founded in 2006, didn't get into news until 2011 (Rieder, 2015a). Best known for cat videos, listicles or numbered lists of the silly and serious, and posts that go viral like #thedress, BuzzFeed's transition into news may be due to a desire to reach Millennials since BuzzFeed's president said, "Traditional media have given up on young people" (Buzzfeed's newest, n.d.; BuzzFeed president, n.d.). Because BuzzFeed's business model is based on links going viral rather than consumers visiting its site or mobile app, Millennials may not be aware of or even care that BuzzFeed mixes up cat videos and silly lists with serious news on its site and app. But because BuzzFeed now wants to be a serious destination for news, it has created a separate app just for news that feels more like NYT Now than BuzzFeed (Isaac, 2014).

HuffPost College

One of many sections of the online-only Huffington Post targeted to specific groups, HuffPost College posts news about happenings on college campuses and issues of interest to and about Millennials. Examples of HuffPost College stories include criticisms of how UC Berkeley handled sexual harassment complaints against a prominent astronomy professor, a poll story that found Millennial women were optimistic about equal pay in the workplace, and underfunding research on gun violence. Stories posted on HuffPost College are written by Huffington Post staff, college students, and picked from the Associated Press and other wire services.

HuffPost is doing a credible job of posting stories about Millennials and of interest to Millennials, but why are these stories segregated in a section called HuffPost College? Many of these stories warrant being included on HuffPost's "Front Page," but they are relegated to a section that not only says Millennials are not important enough to be included with top news stories but also by using "College" in the section label, the online news platform is losing an opportunity to reach Millennials who have graduated college or never attended college.

"Millennials and News" Was Founded by a Journalism Professor and Millennial

The vision of "Millennials and News" is to make news important to Millennial lives. When my Millennial daughter Alexandra Wilson and I co-founded "Millennials and News" on Facebook in 2011, we wanted to encourage news engagement while highlighting three types of news: (1) news Millennials should know, (2) news of interest to Millennials, and (3) news about Millennials. To achieve that goal, every morning, seven days a week, we post news on the "Millennials and News" Facebook page curated from trusted legacy and digital-only news sources such as the *New York Times*, CNN, NPR, the *Washington Post*, the *Los Angeles Times*, *USA Today*, the BBC, and NBC News. Sometimes we post stories from BuzzFeed, the Huffington Post, and USA Today College. Each posted story is introduced with a comment that describes the essence of the news story in the context of why it matters. For news that Millennials should know, we look for the story written with clarity and context so that it can be understood by those who may not have background on the subject.

"Millennials and News" posts top national and international stories, political stories, business stories, personal finance, crime, health news, sports, and lots of technology, especially social media. Trends are extremely important, and with the 2016 presidential election, stories are regularly posted about the candidates and the issues. Important breaking news that Millennials should know at the moment the news breaks is posted with an all-caps BREAKING NEWS lead. Examples of breaking news that has been posted include the Boston Marathon bombing, Vice President Joe Biden's decision not to run for president in 2016, and the death of Harper Lee, the author of the classic must-read book, *To Kill a Mockingbird.*

Celebrity news stories are rare and opinion columns are never posted on "Millennials and News." While "Millennials and News" has succeeded in posting news that Millennials should know and non-celebrity stories of interest to Millennials, it has failed in regularly posting stories *about* Millennials because trusted news sites publish so few stories about Millennials.

As a journalism professor, I use "Millennials and News" in my classes, and at least once a week I require my students to post a comment on a story of interest, and once a week we discuss the stories posted on the Facebook page. The comments and discussion not only provide Millennial insight about the news but they also reinforce the belief that being informed about the news is important. More than a dozen universities around the country use "Millennials and News" in their classrooms.

NYT Now App and the Future of Paying for News

If you disregard the fact that when the mobile app NYT Now was launched in 2014, it had a digital subscription price of $8.00 monthly, it would be a model for what legacy media can do when it's committed to reaching Millennials. NYT Now may have been *New York Times* lite when it launched but it had some really cool features that would appeal to Millennials if they would only pay for it. But apparently Millennials and older mobile consumers didn't want to pay for the app because one year after it was launched, the *New York Times* announced the mobile consumer-targeted news app would be free (Beaujon, 2014). It wasn't surprising that Millennials would not pay $8.00 a month to get news on NYT Now because Millennials also don't pay for the parent brand, according to a Pew Research Center Report. Despite more than fourth-fifths of younger and older Millennials' awareness of the

New York Times brand, less than one-fifth access it for news (Pew Research Center, 2015, p. 18). Or if Millennials do read stories published by the *New York Times*, they are likely reading them for free on Facebook until they hit the monthly maximum quota.

There are multiple explanations as to why Millennials don't pay for news, regardless of the platform. The first and most obvious explanation is children don't grow up paying for news; their parents pay for news. And even if Wave I Millennials had grown up with a print newspaper delivered to the driveway or front door before dawn, they were likely unaware that their parents paid for the newspaper subscription. For Wave II Millennials whose parents may have a digital subscription that allows them to access the news on their laptop or mobile device, there would be even less awareness of paying for news.

Even if after graduation Wave I Millennials wanted to pay for news, which is doubtful, they likely couldn't have afforded to pay for news because when they graduated from college, they were met with a job market severely crippled by the Great Recession of 2008 and many had to begin paying back their student loans. In other words, paying for news would not have been a priority.

As an explanation for why Millennials don't pay for news, it cannot be ignored that Millennials did not come of age paying for anything on the Internet unless they were shopping online. Millennials don't want to pay to download a favorite song. They would rather stream music, TV, and movies with commercials than pay a monthly subscription to stream without commercials. Even when YouTube introduced YouTube Red, a $10-per-month subscription service without ads, seven out of seven Millennials who commented on the *New York Times* story posted on the "Millennials and News" (2015) Facebook page said they would rather watch YouTube videos with ads than pay a subscription fee for ad-free videos (Dougherty & Steel, 2015). And when it comes to news, many Millennials may be philosophically opposed to paying for news on the Internet. When given a five-point scale from 1 to 5, with 1 representing strongly disagree and 5 representing strongly agree, 74% of Wave I and 71% of Wave II Millennials picked a 4 or 5 to indicate a high level of agreement with the statement, "News on the Internet should be free." But Millennials aren't the only generation with this belief about free news on the Internet, according to the News for a Mobile-First Consumer Survey. Seventy percent of Generation X and 72% of Baby Boomers also picked a 4 or 5, endorsing the belief that news should be freely available on the Internet.

Why is the vast majority of each generation against paying for news on the Internet? Except for the online edition of the *Wall Street Journal*, from the beginning news was free on the Internet, that is, until news organizations began erecting paywalls to replace lost display and classified advertising revenue when the newspaper business model collapsed. But as some news organizations quickly discovered, once news is free, it's hard to get the public to pay for it.

While the parent of NYT Now has been very successful in getting 1.1 million people to pay for a digital-only subscription to the *New York Times*, that success did not transfer to the now free NYT Now mobile app (*New York Times*, 2015). As a free mobile app, NYT Now offers far fewer *New York Times* stories than the original paid app. Stories are curated from around the Web, including from *Wired*, *Austin Chronicle*, *Time*, *The Atlantic*, and *The Guardian*. In addition to a mini-crossword puzzle, the app provides briefings and guidance about news that's important and interesting (Sebastian, 2015a). NYT Now does have to tread carefully in its tone so it does not turn off Millennials. The fine line between editorial guidance and editorial paternalism that assumes Millennials aren't sophisticated enough to understand the news should never be crossed.

Snapchat's Discover

The mobile app Snapchat, the latest addition to the social media landscape, created the Discover feature for news organizations to showcase news that appeals to Millennials. Introduced in January 2015, Discover "allows Snapchat users to view news in the form of Snapchat Stories. After 24 hours, the stories disappear and are replaced with fresh content" (Tweedie, 2015).

Snapchat's first news partners in the Discover news initiative included CNN, Vice, Yahoo News, *National Geographic*, the *Daily Mail*, and ESPN. And while Snapchat is extremely popular with Wave II Millennials, Discover is too new to know whether the goal of engaging younger generations and counteracting their "lack of interest in the news" can be achieved (Blum, 2015).

In fact, one Millennial is skeptical about Discover as a vehicle for engaging Millennials with news:

> I don't really see the point. It would never occur to me to get all my news from Snapchat. I think Snapchat as a media platform is interesting. It makes sense in the way that it's all self-destructing so it gets replaced every 24 hours with new news, but I don't see society gravitating to Snapchat to get their news. Especially not anyone who isn't already using Snapchat. (Kosoff, 2015)

Social News Network Integrates Social Media Platforms

Robert Quigley (personal communication, July 20, 2015), a former social media editor at the *Austin American-Statesman* who is now a senior lecturer at the University of Texas at Austin School of Journalism, said in an email that he had two goals in mind when he started Social News Network in 2012: (1) give college students the practical skills needed to either work as a social media specialist, in journalism or outside, or to just more effectively use social media as part of another professional career; (2) provide aggregated news for college students through Twitter, Facebook, Snapchat, Pinterest, and Tumblr. Students post news, sports, and entertainment 15 hours a day throughout the year, even though between semesters, the Social News Network is idle.

TheSkimm Arrives in Millennials' Morning Email

Another free platform created to cultivate Millennial news engagement is theSkimm. Created by two Millennial women for Millennial women, theSkimm, delivers the most important news of the day via an email newsletter Monday through Friday morning (Sebastian, 2015c). Each news story in the emailed newsletter is embedded with links to the original news stories that are the source material for the news. Although the co-founders are former NBC news producers, they write theSkimm using a conversational tone that can be very appealing. Because theSkimm co-founders were always the "go-to source for friends seeking the scoop on current events or breaking news," the co-founders appear to have adopted a persona of informing the uninformed, which sometimes leads to dumbing down the news that some may consider disrespectful to the Millennial women theSkimm was created to appeal to.

TheSkimm also uses a flippant tone in telling the news, which can be engaging when used appropriately, but a flippant tone is not appropriate for every news story. For example, there's no place to be flip when reporting that the Confederate flag was taken down on the South Carolina capitol grounds after a 21-year-old white high school dropout massacred nine African Americans as they attended Bible Study at a historic black church in Charleston, South Carolina. After the gunman was arrested, his anti-black website, which included photos of him posing with a Confederate flag, was found. Here's what theSkimm email, which was delivered to my mailbox at 5:19 a.m., July 10, 2015, said:

THE CONFEDERATE FLAG

THE STORY

Today, South Carolina is taking the Confederate flag down from its capitol grounds. For good.

TELL ME MORE.

The Confederate flag has flown at the state capitol for more than 50 years, ever since the state put it up to protest the civil rights movement. Supporters—who say the flag symbolizes Southern heritage—have always said let it fly. A lot of other people have said … really? But then last month, nine black people were shot and killed at a historic black church in Charleston. Photos later surfaced of the white suspect posing with the Confederate flag. And the rest of the country said, "Are you kidding me with that flag, South Carolina?" Cue SC's gov. saying the flag has to go, followed by major companies dropping their Confederate flag merch. Insert a state lawmaker (who counts the Confederate president as her ancestor) giving this speech. Now, it's coming down. (TheSkimm, 2015)

USA TODAY College

Founded in 2011, USA TODAY College focuses on news that college students are interested in and news happening on college campuses that they should be aware of (Musulin, 2014). The stories are original and reported by a network of USA TODAY College Collegiate Correspondents who apply for this intern-like opportunity that is described on the website:

> USA TODAY College is looking for the best undergraduate student journalists to join the nation's premier multimedia reporting program! We want out-of-the-box thinkers, unconventional storytellers and passionate content creators who will shape the future of the media industry. She/he will have a strong understanding of what the USA TODAY College audience wants—and even the things they don't know they want!
>
> Collegiate correspondents will research, pitch, report and produce one multimedia story per week for USA TODAY College. The program is approximately 17 weeks long with participating students reporting remotely from their respective campuses. Their work will appear on USA TODAY and across Gannett's nationwide network of 93+ media organizations. (USA Today College, n.d.)

USA TODAY College correspondents have reported stories about whether college students think laws allowing carrying guns on college campuses will make them safer (Funke, 2015) and a guide to the 2016 Democratic and Republican presidential candidates, which is useful regardless of generation (Rausch, 2015).

To engage Millennials in news, Millennials and non-Millennials are creating innovative news platforms and so are Fortune 500 companies, entrepreneurs, and journalism professors. Strengthening the bond between Millennials and engaging with news may start with efforts like those listed in Table 5.6, but a more sustained and collaborative plan is needed to make a long-term difference. Suggestions for how to accomplish that will be explored in Chapter 10.

· 6 ·

RACE, ETHNICITY, AND MOBILE

The year 2015 may be remembered as the "Year of Race and Ethnicity." With racial and ethnic flare-ups and setbacks, as well as progress and reconciliation, 2015 was the year race and ethnicity climbed back to the top of the media and public agenda. In fact, in a Gallup Poll taken July 8–12, 2015, race relations and racism were No. 3 on the public agenda behind the economy and dissatisfaction with government/Congress; immigration and undocumented immigrants were No. 4. The economy and dissatisfaction with government/Congress were each mentioned by 13% of poll respondents; 9% said race relations and racism were the most important problem; and 7% singled out immigration and undocumented immigrants (Riffkin, 2015).

The first half of 2015's "Year of Race and Ethnicity" was dominated by news of the deaths of unarmed African American males at the hands of mostly white police officers, including the aftermath of Ferguson, Missouri, in which Michael Brown, an unarmed black teen, was shot and killed by a white police officer. More deaths of unarmed black males followed in Staten Island, New York, North Charleston, South Carolina, Cleveland, Baltimore, and Chicago, among others. But it was the unthinkable racist act in Charleston, South Carolina, that began the summer of 2015 and underscored that the deaths of unarmed black males as a result of excessive police force were not the only race problem in America.

It was June 17 and a 21-year-old white high school dropout attended Bible Study at Emanuel AME Church, a historic black church in Charleston, South Carolina. After participating in the Bible Study, the 21-year-old pulled out a gun and shot and killed nine African American men and women who were also attending the Bible Study. During the investigation of the horrific crime, the gunman's online racist rants and photographs showing him posing with the Confederate flag were found, prompting the governor of South Carolina to call for the removal of the flag from South Carolina's capitol grounds. After debates in the South Carolina senate and house, the state legislature voted to take the Confederate flag down.

The shocking massacre of nine African American men and women during Bible Study at a historic black church by a 21-year-old white gunman forced a national consciousness-raising that, perhaps, hadn't taken place since the nonviolent civil rights protests over a half century before that eventually culminated in the passage and signing into law of the 1964 Civil Rights Act and the 1965 Voting Rights Act. The massacre of black Bible Study participants in Charleston and the subsequent debate about the presence of the Confederate flag on the grounds of South Carolina's capitol building also led to a global conversation on Facebook; it was No. 10 on Facebook's Top 10 list of most-talked-about global topics in 2015 (Cameron & Darwell, 2015). In the U.S., the Charleston massacre was No. 3 on Facebook's Top 10 list of most-talked-about topics, and the Baltimore protests in the aftermath of the death of Freddie Gray, a 25-year-old black male who died in police custody, was No. 5 (Cameron & Darwell, 2015).

With the deaths of unarmed blacks, including an 18-year-old in Ferguson, Missouri, and a 12-year-old in Cleveland at the hands of police, and the #BlackLivesMatter hashtag and movement that provided a national spotlight to focus the national media's attention, it is not surprising that excessive police force moved race relations and racism to the top of the media and public agenda and became the prism by which race problems in the U.S. were framed in the news. While a frame can focus attention, a prism can distort reality, and the reality is a laser-like focus on excessive police force can leave other important news untold, distorting the story of race relations in America. That's why the 1968 Kerner Commission Report remains relevant in 2015, the Year of Race and Ethnicity. The Commission criticized the news media for not showing understanding or appreciation of black culture, thought, or history or reporting on race relations with "wisdom, sensitivity, and expertise" (Report of the National Advisory Commission on Civil Disorders, 1968, p. 384).

Despite the fact that Latinos too have been confronted with excessive police force, the attention that gets on the media and public agenda is minimal compared to the attention given immigration, which is No. 4 on Gallup's Most Important Problem Poll (Riffkin, 2015).

Of course some of the attention given to immigration in 2015 has been due to inflammatory remarks by Donald Trump, a billionaire real estate mogul and reality TV celebrity turned 2016 Republican presidential candidate. Donald Trump not only inflamed negative sentiment against the largest non-white group in the U.S., the billionaire used a prism of illegal immigration and problem people to distort the reality of immigrants and Latinos in the U.S. Immigrants from Mexico "have lots of problems" and they are "bringing those problems to us," Donald Trump said when he launched his campaign for the White House. "They're bringing drugs. They're bringing crime. They're rapists" (Here's Donald Trump, 2015).

Progress and reconciliation may not balance the flare-ups, setbacks, and inflammatory rhetoric, but they are part of the 2015 story about race and ethnicity, and it remains the media's responsibility to also tell those stories, whether about Univision going public, the first prime-time TV program featuring an Asian American family in 21 years, 1,000 Native American youth from around the country attending the White House Tribal Youth Gathering, or the Emmy nominations of two African American women for "lead actress in a drama series." Those news stories must also be told with wisdom, sensitivity, and expertise as well as context (Faughnder & Villarreal, 2015; Caramanica, 2015; Miller, 2015; Lynch & King, 2015).

U.S. Statistics on Race and Ethnicity

Race and ethnicity not only play a role in what is reported in the news, race and ethnicity also are related to news engagement, beliefs about news, and use of mobile devices for news. With a population of almost 319 million that grows more diverse every year, race and ethnicity are as central to the changing media landscape as the mobile transformation that the media landscape is undergoing.

Non-Hispanic whites now represent less than two-thirds (62.6%) of the U.S. population, and that percentage is expected to decline further as non-white groups increase, especially Latinos. In fact, between the 2000 and 2010 U.S. Census, racial and ethnic minorities accounted for 92% of population growth in the U.S. and 56% of that growth came from Hispanics (Minorities, 2011). According to the most recent U.S. Census, Latinos/Hispanics represent 17%, African Americans 13%, Asian Americans 5%, and Native American 1% (Quick facts, n.d.). The 54 million Hispanics in the U.S. may trace their roots to diverse Spanish-speaking countries with different histories, traditions, and customs, but the Spanish language and Catholic religion that connect them transcend differences, making them a force to be reckoned with on every level. Almost two-thirds (64%) of Hispanics in the U.S. are of Mexican background, 9.4% of Puerto Rican background, 3.7% of Cuban background, 3.1% Dominican background, 3.8% Salvadoran, and 2.3% Guatemalan. And 38.3 million, or 74%, of Hispanics 5 and older spoke Spanish at home in 2012.

Latinos are also a force in the Millennial generation because they represent "21% of the overall Millennial population in the U.S." Many Millennial Hispanics are what Univision, the largest Spanish-language broadcaster in the U.S., calls "billennials" because they "increasingly consume their media in English" (Villarreal, 2015). In fact, a "2012 Pew Research study indicated that English is the predominant language in 34% of Latino households, up from 9% in 2011, while Spanish is on the decline. And U.S. Census Bureau projections show that by 2020, one-third of Latinos ages 5 and older will speak only English at home, up from 25% today" (Villarreal, 2015).

Unlike the Latino population, which has the Spanish language and Catholicism uniting them, languages as well as nations, histories, religions, traditions, customs, and political systems distinguish the 19.4 million Asian Americans who trace their roots to China, Vietnam, Korea, Japan, the Philippines, and India. According to the U.S. Census, there are 4.3 million Chinese (excluding Taiwanese), 3.6 million Filipinos, 3.5 million Asian Indians, 1.9 million Vietnamese, 1.8 million Koreans, and 1.4 million Japanese (Asian/Pacific Americans, 2015). With a 90% rate, Asian American households have the "highest Internet penetration of all race and ethnic groups.

The U.S. Census reports that 5.1 million Americans are Native Americans and Alaska Natives and they make up 1.6% of the U.S. population. The smallest racial and ethnic group is also the most diverse, representing 566 federally recognized Indian tribes with different customs, traditions, languages, and histories. One-fifth (22%) of Natives live on reservations or off-reservation trust lands. It is projected that by 2050, 2% of the U.S. population, or 8.6 million, will be Native (American Indian, 2012).

A close look at U.S. population statistics confirms diversity is about race and ethnicity but it's also about much more, including language, history, religion, traditions, customs and culture, and countries of origin. Ironically, mobile devices appear to transcend these differences, that is, when consumers are asked what they primarily use their mobile devices for.

Regardless of racial or ethnic group, the majority are mobile laggards but that is expected to change as Millennials and post-Millennials replace the oldest generations (see Table 6.1). Still, there are racial and ethnic differences among mobile laggards; Asian Americans and Latinos are least likely to be mobile laggards compared to whites and African Americans. Mobile specialists are more likely to be found among Asian Americans, Latinos, and Native Americans than whites or blacks. As recalled from Chapter 1, mobile specialists use their smartphones and tablets to engage in specific activities and are classified accordingly: social media devotees, Internet hunters, gamers, entertainment fans, news seekers, and eclectics, which combined other specialists into a single specialist category. Analyzing mobile specialists by race and ethnicity finds Asian Americans are more likely to be social media devotees and Internet hunters. Native Americans are more likely to be gamers and eclectics. Hispanics are also found among eclectics.

Overall, entertainment fans and news seekers represent a very small percentage of mobile specialists, which is probably why racial and ethnic differences cannot be found. Likewise, mobile-first consumers are a small percentage of mobile types. Only time will tell if the slightly higher percentage of Asian Americans and Latinos among mobile-first consumers will hold in the future.

Table 6.1: Mobile Types by Race and Ethnicity.

	White	Black	Latino	Asian American	Native American
Mobile-first	1%	1%	4%	3%	0%
Mobile specialists	30%	30%	38%	40%	36%
Mobile laggards	69%	69%	58%	57%	64%

News seekers may represent a small percentage of mobile specialists but that does not mean whites, blacks, Latinos, Asian Americans, and Native Americans are not using their mobile devices to engage with news. It's just that news is not what they primarily use their mobile devices for. Table 6.2 shows each racial and ethnic group is getting news and seeking news but not at the level when newspaper reading was called a thoroughly institutionalized behavior.

Table 6.2: News Consumption by Race and Ethnicity.

	White	Black	Latino	Asian American	Native American
Get news 7 days	62%	55%	43%	57%	55%
Seek news 7 days	49%	39%	29%	39%	37%

Not only is there less consumption of news at a time when there are more outlets to get news and more convenient ways to access it than at any time in history, there is a persistent gap between getting and seeking news seven days a week across each and every racial and ethnic group. For whites, the get-seek news gap is 13 percentage points but that gap increases to 18 percentage points for both Asian Americans and Native Americans. Furthermore, less than half of each race and ethnic group seeks news seven days a week and for Latinos, only 29% seek news every day.

There are not only more outlets to get news but with the addition of smartphones and tablets to the media landscape, there are now more platforms to use for getting news. When News for a Mobile-First Consumer Survey participants who got news at least one day per week were asked from what platform they usually accessed news, it was not a surprise that TV topped the list for four of the five racial and ethnic groups. What was a surprise was that in only eight years, smartphones were among the top three platforms used to access news for whites, blacks, Latinos, Asian Americans, and Native Americans (see Table 6.3).

Table 6.3: Platform Used Most to Access News by Race and Ethnicity.

	White	Black	Latino	Asian American	Native American
TV	41%	56%	32%	25%	63%
Laptop	22%	20%	26%	28%	0%
Smartphone	11%	12%	20%	24%	10%
Print newspaper	10%	1%	3%	13%	10%
Tablet	7%	5%	11%	1%	5%
Other, including print magazine	7%	3%	3%	3%	11%
Radio	3%	3%	4%	4%	0%

Some of the similarities among racial and ethnic groups in news platforms used can also be found in type of news consumed often. Across the five racial and ethnic groups, local news is among the top two types of news paid attention to often. Weather is among the top three news topics paid attention to often and national is among the top four (see Table 6.4).

Table 6.4: Racial and Ethnic Comparison of Top 5 News Categories Paid Attention to Often.

White	Black	Latino	Asian American	Native American
Local (71%)	Weather (71%)	Local (68%)	International (55%)	Local (74%)
Weather (69%)	Local (69%)	Weather (55%)	Local (54%)	Weather (63%)
National (61%)	Breaking (64%)	Breaking (54%)	Weather (52%)	Breaking (58%)
Breaking (59%)	National (54%)	National (51%)	National (49%)	National (37%)
International (41%)	Crime (45%)	International (39%)	Breaking (48%)	International (37%)

The main difference among racial and ethnic groups in news paid attention to often is international and crime news. International news is among the top five types of news for whites, Latinos, Asian Americans, and Native Americans, with international news being the No. 1 type of news for 55% of Asian Americans. African Americans are the only racial and ethnic group to have crime in the top 5; 45% of African Americans pay attention to crime news often.

Why are race and ethnicity factors in getting, seeking, and paying attention to certain types of news? Is it due to socialization, how being informed is valued, perceived benefits of news, or something else? To understand race and ethnic differences, statements about socialization as well as the value and benefits of being informed were analyzed by race and ethnicity. Table 6.5 shows that when it comes to how each racial and ethnic group was raised about the importance of being informed, there were no significant differences across the five groups. As few as 3 out of 10 to almost 4 out of 10 of each racial and ethnic group strongly agreed with the statement: "You were raised to believe being informed is important." On the flipside, the majority of each racial and ethnic group—as many as 7 out of 10—did *not* strongly agree they were raised to believe being informed is important. Being raised to believe being informed is important appears to translate into a sense of duty to be informed for some racial and ethnic groups more than others. More blacks (45%) than any other group strongly agreed with the statement: "We all have a duty to keep ourselves informed about news and current events." Asian Americans (31%) were least likely to strongly agree that they had a duty to keep informed.

Table 6.5: Beliefs about Being Informed about News by Race and Ethnicity. % Strongly Agreeing

	White	Black	Latino	Asian American	Native American
Socialization, Duty, and Group Influence:					
Raised to believe being informed is important*	35%	39%	34%	30%	32%
Duty to keep informed	35%	45%	37%	31%	35%
People important to you value keeping up with news	18%	26%	25%	21%	10%
Benefits:					
Enjoy keeping up with news*	35%	42%	36%	32%	20%
Being informed is empowering*	17%	25%	23%	23%	10%
Depend on news to help with daily life	13%	18%	20%	23%	10%
Keeping up with news is a waste of time	5%	6%	13%	11%	0%

*Differences are not statistically significant.

How one is raised and what one believes can be strong influences on news engagement but so can people who are important. When asked to strongly disagree to strongly agree with the statement, "Most of the people who are important to you value keeping up with the news," Blacks and Latinos were more likely than other racial and ethnic groups to strongly agree with the statement. It should be stressed that people who are important may not be limited to peers. Parents, grandparents, teachers, mentors, bosses, or people in the news or on television who are admired from a distance may be in the reference group of important people. Of all the racial and ethnic groups, Native Americans were least likely to strongly agree that people important to them valued keeping up with the news.

Benefits

Socialization may be the impetus behind getting and seeking news but unless there are perceived benefits, socialization may be limited in its influence. That's why benefits of keeping informed of the news were analyzed by race and ethnicity. There were no statistically significant differences in racial and ethnic groups for enjoyment and empowerment. And although more enjoyed being informed than felt empowered by being informed, the majority of whites, African Americans, Latinos, Asian Americans, and Native Americans did *not* strongly agree with the statements: "You enjoy keeping up with the news" and "Being informed of the news makes you feel empowered." Furthermore, less than one-fourth of each racial and ethnic group strongly agreed with the statement, "You depend on the news to help you with your daily life." Asian Americans were most likely and Native Americans were least likely to strongly agree that news is depended on for daily life. And although there was no overwhelming strong agreement with most statements about the benefits of news, only a small percentage (13% or fewer) of whites, African Americans, Latinos, Asian Americans, and Native Americans strongly agreed with the statement, "Keeping up with the news is a waste of time."

How one is socialized about the importance of being informed and the benefits received from keeping up with the news can influence whether one engages with news but so can evaluations about press performance. That's why evaluations of the press, using A through F grades and assessments of specific criteria, were analyzed by racial and ethnic groups.

According to Table 6.6, more African Americans and Asian Americans than other racial and ethnic groups gave the press high grades for its news coverage; 63% of African Americans and 60% Asian Americans graded news coverage A or B compared to 54% of Latinos, 50% of Native Americans, and 44% of whites. When the focus of the grade was on news coverage of the most important problem, about the same percentage of Asian Americans gave the press a grade of A or B but for the other racial and ethnic groups, far fewer gave such high grades. At 25%, Native Americans were least likely to assign a grade of A or B for news coverage of the "most important problem."

Table 6.6: Evaluating News Media Performance by Race and Ethnicity.

	White	Black	Latino	Asian American	Native American
A-B grade of news	44%	63%	54%	60%	50%
A-B grade most important problem news	34%	50%	44%	59%	25%
Press has too much freedom	11%	15%	15%	24%	5%
Press helps protect democracy more than hurt democracy	12%	14%	20%	15%	0%
Watchdog of powerful people and government*	30%	30%	31%	23%	15%
More helpful and caring like good neighbor	19%	30%	28%	25%	5%
More concerned about being first than right*	28%	30%	27%	29%	30%
Most news is biased*	25%	19%	24%	23%	26%
News media care little about people like you	17%	24%	22%	17%	26%
News stories are often inaccurate*	14%	16%	19%	17%	35%
Poor job of providing context	12%	18%	19%	16%	20%
Willing to admit mistakes	7%	13%	17%	12%	0%
News projects accurate representation of constituent groups	7%	13%	16%	19%	0%

*Differences are not statistically significant.

Grades are important but they are not the only way to evaluate press performance, as can be seen in Table 6.6. Another assessment of the press is in the context of its role in society. The press in the U.S. is not just an institution; a free and independent press is central to a democratic society. That's why the authors of the Bill of Rights included a free press among the five freedoms in the First Amendment to the U.S. Constitution. Despite the importance of press freedom to democracy, one-quarter (24%) of Asian Americans strongly agree the "press has too much freedom to do what it wants." And the five racial and ethnic groups are not convinced that "news organizations protect democracy more than hurt democracy." Only one-fifth or less of each group strongly agreed "news organizations protect democracy more than hurt it." Explained another way—at least 80% of whites, African Americans, Hispanics, Asian Americans, and Native Americans do *not* strongly agree that news organizations protect democracy.

Historically, the press's role in society has also been as a watchdog, which is why the racial and ethnic groups were compared on the statement, "It's important for the news media to be a watchdog of powerful people and the government." Results showed no significant differences, with as many as 3 in 10 strongly agreeing on the press's watchdog role. In the last decade, research suggested a more neighborly role for the press, which is why survey participants were asked to strongly disagree to strongly agree with the statement, "The news media should be more helpful and caring like a good neighbor" (Poindexter et al., 2006). An analysis of that statement by race and ethnicity found African Americans (30%), Latinos (28%), and Asian Americans (25%) were significantly more likely than whites (19%) and Native Americans (5%) to strongly agree the press should be more helpful and caring like a good neighbor.

And it may be that belief about the press being more like a good neighbor that's influencing the statement "News media care little about people like you." Native Americans (26%), African Americans (24%), and Latinos (22%) are more likely than whites (17%) and Asian Americans (17%) to strongly agree with the statement: "News media care little about people like you."

Grades and evaluations of the press's role in society capture the public's perception of the press, which may influence news engagement, but they don't assess individual journalism fundamentals, from accuracy to bias. To the discredit of the press, there was no significant difference in the strong agreement of whites, blacks, Latinos, Asian Americans, and Native Americans on the subject of accuracy. Approximately 3 in 10 of each racial and ethnic group strongly agreed with the statement, "News organizations are more concerned about being first than being right." There was also no significant difference in attitudes about press bias or press inaccuracy. Up to 1 in 4 strongly agreed with the statement, "Most news is biased." And while a minority strongly agreed, "News stories are often inaccurate," it's important to emphasize that the vast majority of the racial and ethnic groups did *not* strongly agree with the statement.

Even though it represented a minority of each racial and ethnic group, differences did surface in beliefs about the press's willingness to admit mistakes and providing context to understand a news story. Almost one-fifth (17%) of Latinos strongly agreed the press is "willing to admit mistakes," compared to 0% of Native Americans. When asked if the press does a poor job of providing context to understand a story, one-fifth (20%) of Native Americans, 19% of Latinos, 18% of blacks, and 16% of Asian Americans strongly agreed. Only 12% of whites strongly agreed with the statement.

One of the statements reflected both accuracy and an implied evaluation of journalism's coverage of racial and ethnic groups. The roots of the statement are in the 1947 Hutchins Commission Report that stated there are five requirements the press is expected to meet in a free society. The Hutchins Commission Report, which was primarily funded by *Time Magazine* founder Henry Luce, sought to answer the question: "Is the freedom of the press in danger?" but it became one of those national reports that symbolized expectations about news coverage of race and ethnicity.

Despite having been published in a pre-TV age, the Hutchins Commission Report requirement that the press project a representative picture of the constituent groups in society resonates in today's mobile age. While a vast number of constituent groups exist in the U.S., some journalism scholars interpret this requirement through a race and ethnicity frame, especially in light of research that documents news media's failures to include and accurately represent African Americans, Latinos, Asian Americans, and Native Americans in news coverage (Poindexter et al., 2006; Poindexter, 2009). In fact, some of the language elaborating on this requirement makes it easy to interpret it through a racial and ethnic lens:

> When the images they portray fail to present the social group truly, they tend to pervert judgment. Such failure may occur indirectly and incidentally.
>
> Responsible performance here simply means that the images repeated and emphasized be such as are in total representative of the social group as it is. The truth about any social group, though it should not exclude its weaknesses and vices, includes also recognition of its values, its aspirations, and its common humanity. The Commission holds to the faith that if people are exposed to the inner truth of the life of a particular

> group, they will gradually build up respect for and understanding of it. (A free and responsible press, 1947, pp. 26–27)

How well do racial and ethnic constituent groups think the news media projects a representative picture? An analysis of that statement in the News for a Mobile-First Consumer Survey suggests that for most racial and ethnic groups, the press comes up short in that evaluation. In fact, less than one-fifth of each racial and ethnic group strongly agreed that the press "projects an accurate representation of the constituents groups in U.S. society" and no Native Americans strongly agreed with that statement.

Trust

Negative evaluations of press performance have the potential to take a toll on press trust, especially evaluations related to bias and accuracy. And it appears that journalists have been negatively affected more than the news organizations they work for. Furthermore, the trust evaluations do not differ significantly across racial and ethnic groups. Table 6.7 shows a larger percentage of whites, blacks, Latinos, Asian Americans, and Native Americans trust local TV news over all other news sources. Following local TV news in trust are the local newspaper, both print and online, network TV news, family and friends, and national newspapers.

Cable news is not trusted the same across all racial and ethnic groups. More Latinos than any other racial and ethnic group trust cable news a lot, while more Asian Americans than any other racial and ethnic group trust Google, journalists, Facebook, government websites, Twitter, and bloggers.

Table 6.7: Sources Trusted for News by Race and Ethnicity.
% Saying A lot

	White	Black	Latino	Asian American	Native American
Local TV news*	42%	44%	47%	41%	35%
Local newspaper online and print*	34%	34%	37%	35%	10%
Network TV news*	30%	28%	30%	35%	20%
Friends and family*	28%	27%	27%	40%	35%

Table 6.7: (*continued*)

	White	Black	Latino	Asian American	Native American
National newspaper*	24%	28%	30%	33%	10%
Cable news	24%	30%	34%	31%	5%
Google	18%	23%	31%	37%	20%
Journalists	15%	20%	20%	24%	5%
Facebook	12%	11%	21%	25%	10%
Govt. websites	10%	16%	16%	21%	5%
Twitter	6%	10%	15%	20%	0%
Bloggers	6%	9%	13%	20%	10%

*Differences are not statistically significant.

Mobile News

While it's too early to know the full effect the mobile transformation of the news media landscape will have on consumer news engagement, evaluations of news media performance, and trust of the press, it's *not* too early to establish a baseline for racial and ethnic mobile news use. In fact, it's imperative that a baseline be established now because past legacy news use patterns appear not to apply in today's mobile news media landscape.

As a mobile device, smartphones are more popular than tablets for all racial and ethnic groups and more Latinos and Asian Americans own smartphones than whites, blacks, and Native Americans. According to Table 6.8, a whopping 80% of Latinos and Asian Americans own smartphones compared to 58% of whites, 59% of blacks, and 55% of Native Americans.

Engaging with news may not be what whites, African Americans, Latinos, Asian Americans, and Native Americans primarily use their mobile devices for, but that doesn't mean they don't use their mobile devices for news. Of the five racial and ethnic groups, Native Americans (73%) followed by Asian Americans (69%), Latinos (57%), African Americans (45%), and whites (39%) use a mobile device to engage with news.

And while from 39% to 73% of racial and ethnic groups use their mobile devices to engage with news, it's important to emphasize that everyone does not use their mobile device for news. Sixty-one percent of whites, 55% of blacks, 43% of Latinos, 31% of Asian Americans, and 27% of Native Americans do not use their mobile device for news.

Table 6.8: Mobile News Engagement and Attitudes by Race and Ethnicity.

	White	Black	Latino	Asian American	Native American
Own smartphone	58%	59%	80%	80%	55%
Own tablet	46%	42%	66%	69%	45%
Use mobile device to engage with news	39%	45%	57%	69%	73%
Top 2 methods used to engage with news on mobile	App 40% Browser 21%	App 28% Facebook 27%	App 35% Browser 21%	Facebook 33% App 29%	App 64% Facebook 27%
0 news apps	42%	33%	27%	23%	27%
3 or more news apps	25%	24%	33%	33%	13%
Strongly agree easy to keep up with news on mobile	24%	33%	34%	32%	25%
Strongly agree frustrating to read news on mobile	15%	7%	17%	19%	10%
A-B Grade for presentation of news on mobile	65%	81%	78%	73%	64%

Because there are at least a half dozen ways to access news on a smartphone or tablet, the five racial and ethnic groups were analyzed to determine which method they used to access news on a mobile device: app, browser, social media, notification, breaking news alert, search engine, or other. Results of the analysis showed apps and browsers were the top two methods for mobile news engagement by whites and Latinos, but for African Americans and Native Americans, apps and Facebook were the top two methods to engage with news. Asian Americans also used apps and Facebook but Facebook was preferred over apps for engaging with news.

Despite the popularity of news apps, a segment of each racial and ethnic group has no news apps on their mobile devices. More than two-fifths (42%) of whites have no news apps; one-third (33%) of African Americans have no news apps. Over a quarter of Latinos (27%) and Native Americans (27%) have no news apps and 23% of Asian Americans do not have news apps on their mobile devices. While a segment of each racial and ethnic group has no news apps, there's also a segment with three or more news apps downloaded on their mobile devices: Latinos (33%); Asian Americans (33%); whites (25%); African Americans (24%); Native American (13%).

Despite the fact that there are holdouts in using mobile devices for news, there are positive indicators that will change in the future, especially because a quarter to one-third of each racial and ethnic group strongly agrees it is easy to keep up with news on mobile and less than one-fifth finds it frustrating to read news on mobile. Perhaps, the best indicator for future engagement with news on mobile is that at least three-fifths and as many as four-fifths give a grade of A or B to the presentation of news on mobile. In fact, a larger percentage, from 13% to 24%, give a grade of A-B to mobile news than news in general.

Social Media

Social media are as much a part of today's news landscape as are mobile devices. Social media can both facilitate news engagement and serve as a barrier to news engagement because non-news activity on social networking sites can eat up time and attention that might otherwise be devoted to getting news. Still, the fact that Facebook is among the top two ways consumers access news on mobile suggests that social media are facilitating news engagement.

Facebook is not just a vehicle for each racial and ethnic group to get news on mobile; Facebook appears to be a dominant part of the lives of each racial and ethnic group. According to Table 6.9, three-quarters of whites, blacks, and Native Americans and more than four-fifths of Latinos and Asian Americans are on Facebook. The majority of each racial and ethnic group also has a Facebook mobile app that makes it convenient to both do non-news things and access news anytime and from anywhere. And each racial and ethnic group is using Facebook time for news, with a larger percentage of Asian Americans, compared to the other racial and ethnic groups, using Facebook for news; 45% of Asian Americans, 34% of Latinos, 25% of blacks, 19% of whites, and 13% of Native Americans devote one-third or more of Facebook time to news.

Instagram is not as popular as its Facebook owner but for Asian Americans, the photo and video-sharing site is slightly more popular than Twitter, according to Table 6.9. And although Twitter also lags behind Facebook in the social media arena as measured by percentage of users, Twitter is more organically connected to news because both journalists and citizen journalists use it to tweet the news and mobile devices make it convenient for users to follow news anytime. Plus, journalists' Twitter handles, hashtags tied to specific news stories, and trending topics can make the news experience on Twitter more engaging than the news experience on other social media, legacy, and online-only platforms.

Twitter also falls short on mobile apps when compared with Facebook. This, of course, is not surprising since there are far fewer Twitter users than users of Facebook. But Twitter is competitive with Facebook on time spent on news, with about the same percentage of blacks, Latinos, Asian Americans, and Native Americans spending at least one-third of social media time on news. However, a larger percentage of whites on Twitter than Facebook spend at least one-third of their social media time on news (28% vs. 19%).

Just as a larger percentage of Asian Americans than any other racial or ethnic group devotes at least one-third of Facebook time to news, a larger percentage of Asian Americans use at least one-third of their Twitter time engaging with news. As Table 6.9 shows, almost half (48%) of Asian Americans compared to less than one-third (31%) of Latinos, 28% of whites, one-quarter (24%) of blacks, and 17% of Native Americans devote at least one-third of Twitter time to news.

The differences among racial and ethnic groups are not statistically different but the Twitter news engagement activities—following a journalist, clicking and retweeting news, and clicking trending news—are meaningful, nonetheless. Though it's not clear whether Twitter users follow journalists or news organizations because it's easy, they are trusted, or for some other reason, it is clear that following journalists or news organizations is a popular news engagement activity on Twitter and all racial and ethnic groups do it.

While a larger percentage of Hispanics and Asian Americans than African Americans are on Twitter and have a Twitter app and there are no significant differences across racial and ethnic groups in engaging with news on Twitter, it is important to note that African Americans have carved out a unique space in the Twitter universe that has come to be known as "Black Twitter" (Brock, 2012). Calling Black Twitter a "force," *Atlantic* writer Donovan Ramsey observes:

> This was the network largely responsible for focusing the nation's attention to the killing of Mike Brown in Ferguson, Missouri, last August. Witnesses to Brown's killing broke the news via social media. Within moments, their accounts of what happened spread through the Twittersphere with the hashtags #Ferguson and #MikeBrown. (Ramsey, 2015)

Table 6.9: Social Media, News, and Race and Ethnicity.

	White	Black	Latino	Asian American	Native American
On Facebook	73%	75%	82%	86%	75%
Have Facebook app	56%	67%	72%	76%	67%
1/3-plus Facebook time on news	19%	25%	34%	45%	13%
On Instagram	20%	25%	42%	50%	10%
Have Instagram app	22%	30%	42%	51%	13%
On Twitter	30%	34%	48%	47%	30%
Have Twitter app	27%	29%	42%	45%	27%
1/3-plus Twitter time on news	28%	24%	31%	48%	17%
Follow journalist or news organization on Twitter*	39%	37%	48%	42%	33%
Often click tweeted news story*	22%	19%	25%	18%	0%
Often click trending news story*	21%	18%	26%	15%	17%
Often retweet news story*	11%	15%	18%	10%	0%
Depend on social media for daily life	9%	12%	16%	23%	10%

*Differences are not statistically significant.

Black Twitter, which has been the subject of academic research, news media reports, and even the focus of a Wikipedia page, has been defined by Black Twitter expert Meredith Clark, a journalism professor at University of North Texas, as a "temporally linked group of connectors that share culture, language and interest in specific issues and talking about specific topics with a black frame of reference" (Ramsey, 2015). In her Ph.D. dissertation, Clark (2014) said that while the definition of Black Twitter is "ever-shifting," there are some consistent attributes:

> It can be described as a phenomenon—the creation of Twitter trends through conversation, the collective experience of watching a televised event, or participating in an online chat. It can also be described as a community—a structure that Twitter

> users feel they are a part of, and willingly participate in as a form of maintaining membership. It has been defined as a medium for cultural conversation—a tool through which anyone (regardless of race or identity) who understands the historic roots of the slavery and Jim Crow narratives, Black entertainment and folklore and contemporary cultural practices can participate in order to exchange ideas, develop opinion leadership on a topic, or simply learn more. (Clark, 2014, p. 31)

Black Twitter's ability to influence public opinion and bring awareness of an issue to the attention of the news media has no doubt played a significant role in the return of race relations and racism to the top of the media and public agenda in 2015, as evidenced by a Gallup Poll taken July 8–12, 2015 (Riffkin, 2015). If Black Twitter can influence the news media's agenda, how can the news media influence Black Twitter's agenda and get more blacks engaging with news, especially because mobile devices are always accessible? That's an issue worth exploring as part of an overall effort to make news more of a priority and better able to compete with the social media on the screens of mobile devices.

Depending on Social Media to Help with Daily Life

Social media are woven into the lives of consumers in ways that could not have been imagined a decade ago. But does that mean consumers depend on social media for help with daily life? Table 6.9 suggests not. When News for a Mobile-First Consumer Survey participants were asked to strongly disagree or strongly agree with the statement, "You depend on social media to help you with your daily life," the vast majority of each racial and ethnic group did not strongly agree with that statement. At 23%, Asian Americans were most in strong agreement compared to 16% of Latinos, 12% of African Americans, 10% of Native Americans, and 9% of whites who strongly agreed that they depended on social media to help with daily life. As documented in Table 6.5, depending on news to help with daily life was also low but slightly higher than social media.

While it is somewhat ironic that with the high level of participation on social media, especially Facebook, the racial and ethnic groups do not view social media as helping with daily life, this should not be interpreted as social media not having an effect on daily life, especially because of the amount of time and attention devoted to social media. The time and attention devoted to social media mean less time and attention devoted to news, despite its availability on mobile devices anytime and from wherever one may be. With the popularity of social media in the mobile landscape as well as the growing population of people of color and declining non-Hispanic white population, it is more important than ever for news organizations to improve news coverage of racial and ethnic groups and make increasing news engagement among them a high priority. Today's Millennial Generation is the most diverse ever and that diversity will only increase in the post-Millennial Generation, providing even more challenges for news organizations and journalists. How news organizations can address these diversity issues that make the mobile landscape, populated with social media, even more challenging, will be explored in Chapter 10.

· 7 ·

WOMEN AND MEN

Divided and Connected in the Mobile Space

In 2008 the editors of *Women, Men, and News: Divided and Disconnected in the News Media Landscape* and its contributors explored the gender gap in news consumption in the U.S. and around the world with the goal of identifying causes behind the gap and solutions that news organizations, journalism schools, and society might employ to close the gender divide in news consumption, increase news engagement overall, and strengthen the belief that being informed is important. Because *Women, Men, and News* was published one year after the iPhone was unveiled and two years before the introduction of the iPad, the book's editors and contributors could not have foreseen the impact of smart mobile devices or the role they might play in the news consumption habits of men and women. Would smartphones and tablets close the gender consumption divide or enlarge it?

Because mobile devices make it more convenient for men and women to access news anytime and from anywhere and because there are so many news choices available through one's smartphone or tablet, one might expect the gender divide identified in a 1990 pre-mobile world would not exist today. But if the pre-mobile world causes of the gender gap in news consumption are still present, the gender gap in news consumption could very well have persisted or even worsened.

What then has been the impact of mobile devices on the gender gap in news consumption? Are women and men engaged with news at the same level, or are women less engaged with news than men despite the fact that there are more women (161 million) than men (156.1 million) in the U.S. (Women's history month, 2015)? And what impact, if any, has the changing of the generational guard had on the news consumption gender gap? Are Millennial females engaging with news at the same level as Millennial males in today's mobile world, or has the gender gap identified in a pre-mobile world taken hold in the largest living generation?

As *Women, Men, and News* underscored, the pre-mobile news consumption gender gap was not isolated to the U.S. A gender news consumption gap could be found around the world, including the United Kingdom, Germany, Spain, Turkey, Egypt and the Middle East, Nigeria, India, China, Taiwan, South Korea, Brazil, Chile, Mexico, and the Caribbean (Poindexter, et al., 2008). And although some gender gaps have been "fixed" by laws that have made it illegal to discriminate against women in voting; the workplace; universities, including college athletics; and society as a whole, other gender gaps have persisted, despite laws.

Society's gender divide can be found in earnings, job promotions, and industries such as technology and film. A gender gap is also visible in corporate suites, boardrooms, and newsrooms, as well as among world leaders, state and local elected officials, university presidents, news bylines, and the STEM disciplines. For example, U.S. women earn 78 cents for every dollar their male counterpart earns and males' median annual earnings are $50,033 compared to $39,157 for women (Women's history month, 2015).

To fully understand if the news consumption gender gap explored in *Women, Men, and News* (Poindexter, et al., 2008) has persisted, mobile, as well as news consumption patterns and attitudes, will be analyzed in the context of the generations in which male and female news consumers were born.

Mobile Types and Gender

According to Table 7.1, both gender and generation matter in determining if there are gender gaps among mobile-first consumers, mobile specialists, and mobile laggards. Looking at only generation or gender individually would mask the presence of a gender gap. For example, overall, both Baby Boomer and Generation X males and females are more likely than Millennial males and females to be mobile laggards, but within the generations, a slightly different picture emerges: a mobile laggard gender gap exists among Millennials but not among older Baby Boomers and Generation Xers. Millennial females (54%) are more likely than Millennial males (47%) to be mobile laggards but about the same percentage of Baby Boomer and Generation X males (72%) and females (70%) are mobile laggards.

A similar pattern exists among mobile specialists, which includes social media devotees, Internet hunters, entertainment fans, gamers, news seekers, and eclectic mobile specialists. While overall more Millennial males and females than Baby Boomer and Generation X males and females are mobile specialists, about the same percentage of older males (27%) and females (28%) are mobile specialists, but among Millennials, males (48%) are more likely than females (42%) to be mobile specialists.

At this stage of mobile device adoption, there are too few mobile-first consumers to make statistical claims with any certainty but it appears there is no gender gap between Millennial males and females. Only after more consumers become mobile-first will it be possible to determine if this gender-neutral pattern holds.

Table 7.1: Mobile Types by Gender and Generation.

	Baby Boomer and Generation X Males	Baby Boomer and Generation X Females	Millennial Males	Millennial Females
Mobile-first	1%	2%	4%	4%
Mobile specialists	27%	28%	48%	42%
Mobile laggards	72%	70%	47%	54%

Gender Gap in Getting and Seeking News

During the legacy news media era, news didn't find consumers, consumers found news. If consumers wanted news, they had to be proactive by subscribing to a newspaper or news magazine or buying a single copy of a newspaper or magazine at a newsstand. Consumers could also find news by turning on the TV to local or network news at designated times or turning to the 24-hour cable news channels. Today, news finds consumers whether they want it or not. In fact, it is virtually impossible to avoid getting news because it is everywhere, including social media and smartphone and tablet notifications. That's why when consumers are asked if they get news, their news consumption response may be somewhat inflated. But if consumers are asked about seeking news, their responses more accurately reflect their news consumption behavior.

Historically, older generations have gotten news more than younger generations so it is not a surprise that Baby Boomers and Generation X get and seek news more than Millennials (see Table 7.2). And based on the news consumption gender gap examined in *Women, Men, and News*, the 9-point gender gap in getting news daily and 12-point gender gap in seeking news daily among Baby Boomers and Generation X is not a surprise. What is somewhat of a surprise is the gender gap among Millennial males and females. The 11-point Millennial gender gap in getting news daily narrows to a 5-point gender gap in seeking news, yet the pattern of females lagging behind males in news consumption in a pre-mobile environment persists in the mobile news landscape among the youngest generation.

Table 7.2: News Engagement by Gender and Generation.
% Getting and Seeking News Daily

	Baby Boomer and Generation X Males	Baby Boomer and Generation X Females	Millennial Males	Millennial Females
Get news daily	68%	59%	44%	33%
Seek news daily	55%	43%	28%	23%

Gender Differences in Platform Used to Access News

The gender gap is not restricted to getting and seeking news; the gap is also present in the platforms consumers use to get news. Table 7.3 reminds us TV is still the dominant platform for news for Baby Boomers and Generation X but TV is No. 3 on a list of five platforms that Millennials use to access news. Mobile devices are the No. 1 platform Millennials use to access news, followed by laptops and TV. It is interesting that Millennial males are more likely than Millennial females to use laptops for news while the reverse is true for TV. But when it comes to smartphones or tablets, both Millennial males and females use that platform for news about the same. The biggest loser as a platform for accessing news is print newspapers. Once among the top two platforms for accessing news, print newspapers are second from the bottom of the list of platforms, just above radio, which is primarily a platform for music and not news.

Table 7.3: Platform Used to Access News by Gender and Generation.

	Baby Boomer and Generation X Males	Baby Boomer and Generation X Females	Millennial Males	Millennial Females
TV	48%	46%	21%	25%
Laptop	20%	19%	35%	29%
Smartphone or tablet	13%	17%	38%	37%
Print newspaper	9%	8%	2%	4%
Radio	4%	3%	1%	2%
Other*	6%	6%	1%	3%

*Other includes e-reader and print magazine.

Gender Gap in News Topics

While Tables 7.2 and 7.3 provide insight into overall news consumption patterns and platforms, they do not shed light on news topics that consumers pay attention to, and they don't explain the gendered hierarchy of hard news and soft news that legacy newspapers used to distinguish more important from less important topics.

> Traditionally, the distinction between hard and soft news in the newsroom has symbolized a news hierarchy: hard news, at the top of the newsroom ladder, has been viewed as important news that men do; soft news, several rungs below hard news, has been perceived as less important news that women do. (Poindexter & Harp, 2008)

Journalism educators and journalism professionals have tried to distinguish hard news from soft by specifying topics and even writing styles. Generally, public affairs, social problems, crime, government and business actions, as well as accidents have been classified as hard news; soft news has included trends, personalities, lifestyles, family, food, fashion, furnishings, plays, movies, books, celebrity, and health. While not all of the topics listed in Table 7.4 would fit neatly into traditional hard and soft news categories and some news topics, like technology, are a function of the digital, social media, and mobile transformation of the news landscape, there is evidence that the gender divide can be found in hard news and soft news but the gender gap in hard news is not always in the expected direction. For example, both local news and crime news would be categorized as hard news but Baby Boomer, Generation X, and Millennial females are more likely than males in the same generations to pay attention to these two news topics often. A careful examination shows the gender divide for these two news topics is greatest among Millennials; 15 points separate Millennial females from males in paying attention to local news, and there's a 13-point gender difference in crime news.

For other types of hard news—national, international, politics, government, election news, and business news—a news consumption gender gap exists among older generations but for Millennials, the gender gap narrows considerably for national, international, election, and business news.

The gender legacy of soft news in which news about celebrity and entertainment or health was consumed more by females than males persists in the mobile news landscape. Baby Boomer, Generation X, and Millennial females are more likely than their male counterparts to often pay attention to news about health, celebrity, and entertainment. In fact, the 12-point gender gap for attention Baby Boomers and GenXers pay to celebrity and entertainment news widens to a 17-point gender differential for Millennials: 37% of Millennial females compared to 20% of Millennial males often pay attention to celebrity and entertainment news.

A gender gap can also be found when weather and breaking news, two of the five most popular news topics, are analyzed. For both older and younger generations, females are slightly more likely than males to pay attention to weather often. A larger gender gap exists for breaking news; 67% of Baby Boomer and Generation X females, compared to 60% of males in that generational cohort, pay attention to breaking news often. Among Millennials, females (49%) are more likely than males (41%) to pay attention to breaking news.

Technology news may not be the most popular but it is now a must-report news topic for legacy and native online news organizations. In fact, of 11 mobile news apps that I examined, including two native digital, cable, radio, TV, and a newswire, all 11 had a separate section devoted to technology. News about technology is rarely technical; it's about social media and tech companies, including polices, economic successes and failures, and most recently about their employment diversity. Ironically, the leading tech companies lag in racial, ethnic, and gender diversity. Despite high-profile female executives at Facebook and Yahoo, major technology companies have a small percentage of female employees in the workforce, including the executive suite. Not only is there a gender gap in technology companies, there is a gender gap in technology news consumption (Hu, 2014). Plus, the gender gap is similar for Baby Boomers, Generation X, and Millennials. Twenty-seven percent of Baby Boomer and Generation X males read technology news often compared to 16% of their female counterparts. Among Millennials, 33% of males read technology news often compared to 23% of females.

Table 7.4: News Topics by Gender and Generation.
% Paying Attention to Often

	Baby Boomer & Generation X Males	Baby Boomer & Generation X Females	Millennial Males	Millennial Females
Local news	73%	77%	45%	60%
National news	71%	52%	43%	41%
Weather	68%	72%	51%	54%
Breaking news	60%	67%	41%	49%
International news	51%	36%	34%	30%
Sports	48%	16%	36%	17%
Politics	43%	25%	27%	18%
Government	43%	26%	30%	21%
Natural disaster	42%	40%	24%	30%
Election news	38%	22%	20%	18%
Crime	33%	40%	21%	34%
Business news	29%	13%	17%	12%
Technology	27%	16%	33%	23%
Consumer news & personal finance	24%	18%	15%	15%
Accidents	23%	33%	21%	31%
Health news	20%	26%	17%	23%
Environment*	20%	21%	19%	19%
Celebrity & entertainment	12%	24%	20%	37%
Art & culture*	11%	13%	18%	16%

*Differences are not statistically significant.

Mobile News and Gender

As reported earlier in Table 7.3, when participants were given a list of platforms that could be used to access news, mobile devices were the most popular for Millennials, with males and females using smartphones or tablets for news at about the same rate (38% vs. 37%). But for older generations, mobile devices were the third most popular platform for news, behind TV and laptops, with Baby Boomer and Generation X females (17%) slightly more likely than Baby Boomer and Generation X males (13%) to prefer a mobile platform for news. However, when a direct question is asked about using a mobile device to engage with news, Baby Boomers and GenX males use a mobile device to engage with news more than their female counterparts (43% vs. 37%). And while Millennials are much more likely than older generations to use mobile devices for news, there is still a slight gender gap, with 66% of males compared to 61% of females using a mobile device to engage with news.

To provide greater insight into the use of mobile devices for news, survey participants were asked: "From 0% to 100%, approximately what percent of the time that you spend on your mobile device on a typical day is devoted to engaging with news?" The results showed both generational and gender differences. Overall, more Millennial males and females than older Baby Boomer and Generation X males and females devote at least 20% of their mobile device time to news; however, when Millennial males and females are compared directly, a 15-point gender gap emerges. In other words, 70% of Millennial males compared to 55% of Millennial females spend 20% or more of their mobile device time on news.

Multiple Methods for Engaging with News on Mobile Devices

Just as there are multiple platforms to access news, there are multiple ways to engage with news on a mobile device, from apps and browsers to search engines and social media and alerts. More males and females of the older and younger generations use apps to access news on mobile devices than any other method (see Table 7.5), and there are gender differences in methods used to access news on mobile. Baby Boomer and Generation X males are significantly more likely than their female counterparts to use apps for news, but males in the older generations are significantly less likely than females to access news on a mobile device using Facebook.

Table 7.5: Method Used to Engage with News on Mobile by Gender and Generation.

	Baby Boomer & Generation X Males	Baby Boomer & Generation X Females	Millennial Males	Millennial Females
App	46%	32%	32%	35%
Browser	19%	19%	23%	21%
Facebook	12%	22%	22%	24%
Breaking news alert	9%	12%	7%	7%
Search engine	9%	7%	5%	6%
Twitter	3%	2%	8%	4%
Notification	1%	2%	5%	1%
Other	1%	4%	0%	1%

The preference for news apps to engage with news on smartphones does not mean mobile device owners download an unlimited number of apps. In fact, the response to the question, "Approximately how many apps on your mobile device are for news?" found that mobile device owners had a mean of 2.4 news apps and most had no news apps. Plus, as Table 7.6 documents, there is a news app gender gap. Baby Boomer and Generation X females (42%) were more likely than their male counterparts (37%) to have "0" news apps and there was an even larger gender gap among Millennials. Almost one-third (31%) of Millennial females had no news apps on their mobile devices compared to almost one-fifth (18%) of Millennial males.

For three or more news apps, the gender gap flipped for older and younger generations; 32% of Baby Boomer and Generation X males had three or more news apps compared to 24% of their female counterparts. A similar gender gap was found among Millennials; 34% of Millennial males had three or more news apps compared to 23% of Millennial females.

Table 7.6: Number of News Apps by Gender and Generation.

	Baby Boomer & Generation X Males	Baby Boomer & Generation X Females	Millennial Males	Millennial Females
0	37%	42%	18%	31%
1–2	31%	35%	48%	47%
3 or More	32%	24%	34%	23%

While there has been an overwhelming adoption of mobile devices across generations and genders, mobile devices are not yet the dominant platform for engaging with news, and the news app innovation has yet to be adopted by the majority of every generation and gender. Perhaps that's because only 23% to 39% strongly agree that mobile devices make it easy to keep up with news. But those numbers are expected to increase especially because whether male or female or Millennial, GenX, or Baby Boomer, an average of 72% of mobile device owners who use smartphones or tablets give a grade of A or B to the presentation of news on a mobile device.

Social Media as a Mobile Gateway to News

The role that social media play in engaging with news on mobile devices comes into focus in Table 7.5. Facebook is the second most popular method for engaging with news for Baby Boomer, Generation X, and Millennial females. And when Facebook and Twitter are combined, social media become the second most popular method for engaging with news on mobile devices for Millennial males. Just as virtually every news organization produces news apps to make it quick and easy to engage with their news on mobile devices, social media companies have apps for easy, anytime access to their social networking sites. While a mobile social media app can be a gateway to news through social media, it can also be a barrier to engaging with news, especially for Millennials.

> While social media can be used for professional networking and as a platform for breaking news and communicating marketing and political messages, it's the *social* in social media that has turned upside down our understanding about the factors facilitating and impeding engagement with news.
>
> But it's not social media alone that's affecting news engagement; it's the unprecedented anytime, anywhere access to social media through mobile devices such as smartphones, tablet computers, and laptops. Whether in a coffeehouse, riding in a car or on a subway, waiting at the airport, even sitting in a college classroom, connecting to and interacting with social media is easy and quick. That's what makes social media such a potentially powerful impediment to news engagement, especially for Millennials. Social media not only eats up time; it can consume attention and render traditional socializing agents ineffective when it comes to engagement with news. (Poindexter, 2012, p. 18)

Plus, among mobile specialists, more Millennial males and females, as well as Baby Boomer and Generation X females, are social media devotees than news enthusiasts, which makes it less likely that they will pick news over social media on their mobile devices. And since Millennial females are at least two and a half times more likely to be social media devotees than Millennial males and older males and females, getting Millennial females to spend time with news instead of social media is a long-term challenge, especially because social media apps make it so convenient to access social networking sites wherever they are.

Table 7.7 shows how social media apps dominate mobile screens, and for Millennial males and females, Facebook and YouTube are the most popular. While about the same percentage of Millennial males and females have a mobile app for Facebook, YouTube, and LinkedIn, Millennial males are slightly more likely than Millennial females to have a Twitter app, and Millennial females are significantly more likely than Millennial males to have an Instagram mobile app (63% vs. 43%).

Table 7.7: Social Media Apps on Mobile Devices by Gender and Generation.

	Baby Boomer & Generation X Males	Baby Boomer & Generation X Females	Millennial Males	Millennial Females
Have Facebook app	53%	62%	76%	79%
Have YouTube app	51%	50%	79%	79%
Have Twitter app	26%	25%	52%	48%
Have LinkedIn app*	22%	15%	18%	18%
Have Instagram app	15%	21%	43%	63%

*Differences are not statistically significant.

News can be posted, shared, and consumed on all social media apps but Twitter has evolved as the social networking site most associated with news. Millennials go to Twitter to find the latest news, and journalists and news organizations use Twitter to distribute developing news stories. And because politicians and other newsmakers use Twitter to dispense quotable tweets, journalists also rely on Twitter as an on-the-record source for their news stories. Twitter has also enabled journalists to expand their toolbox by live tweeting breaking news and public events and connecting to public issues, concerns, and interests by tapping into what's trending, which frequently ends up as a story on the evening news.

While some of the trending stories that end up on the national news are silly, such as the color of a dress, Twitter has served an important service in helping verify information or correcting the record—such as in the case of the Boston Marathon Bombing—or bringing newsworthy issues to the attention of journalists as it did when an unarmed black teen was shot and killed by a white police officer in Ferguson, Missouri.

The news connection with Twitter is not something Twitter planned when it started up but, nevertheless, it is has become a multifunctional mobile news platform for news consumers, journalists, and news organizations. Twitter's new "Moments" tab should make it easier for consumers to find news and for journalists to monitor what's trending (About moments, n.d.).

Just as there is a gender gap in getting and seeking news, there is a gender gap on Twitter when it comes to news. While, overall, Millennials use Twitter for news more than older generations, a larger percentage of Baby Boomer and Generation X males than females spend at least one-third of their Twitter time on news (28% vs. 20%) and more Millennial males (46%) than Millennial females (34%) spend at least one-third of their Twitter time on news. And although Twitter is more associated with news than Facebook, Table 7.8 underscores that for older and younger females, Facebook is at least equal to Twitter as a platform for engaging with news.

Table 7.8: News Activity on Facebook and Twitter by Gender and Generation.

	Baby Boomer & Generation X Males	Baby Boomer & Generation X Females	Millennial Males	Millennial Females
1/3-plus Facebook time on news	18%	21%	39%	37%
1/3-plus Twitter time on news	28%	20%	46%	34%
Follow journalist or news organization on Twitter	42%	37%	57%	39%
Often click tweeted news	21%	19%	35%	19%
Often click trending news*	15%	20%	32%	24%
Often retweet news	12%	8%	26%	11%

*Differences are not statistically significant.

There are a variety of ways to engage with news on Twitter and Table 7.8 shows engagement with news via Twitter varies by gender and generation. For older and younger generations, more males than females follow journalists or news organizations on Twitter but the widest gender gap is between Millennial males and females. A 15 to 18-percentage-point gap separates Millennial males and females in frequently following journalists or news organizations on Twitter, clicking tweeted news, and retweeting news. While the generational and gender differences in clicking trending news often are not statistically significant, it is noteworthy that an average of one-fifth of Baby Boomers, GenX, and Millennial males and females click trending news often.

The pre-mobile news consumption gender gap examined in *Women, Men, and News: Divided and Disconnected in the News Media Landscape* (Poindexter et al., 2008) has not only emerged in the mobile news landscape, it has woven its way into social media, an innovation of the Millennial generation that has become a platform for getting news that competes with both legacy and native digital news for the time and attention of men and women of all generations. Mobile devices could have and should have helped close the news consumption gender gap, but instead of disappearing in a mobile-transformed news landscape, the news consumption gender gap appears to have gained momentum, expanding into areas that were in their infancy in 2008. Socialization and attitudes about being informed and press performance may offer some clues as to why the news consumption gender gap did not vanish but asserted itself in new mobile and social media incarnations.

Socialization and Civic Responsibility to Be Informed

The socialization process suggests children can be groomed to grow up into future news consumers who value the importance of being informed.

> What children learn through the socialization process is a product of a variety of factors, including what they are exposed to at home, at school, in the park, in the mall, in a religious or cultural setting, or through the media. In any one of these environments, parents, siblings, teachers, peers, religious leaders, politicians, celebrities, athletes, or textbooks, to give a few examples of socialization agents, can directly or indirectly influence exposure to news as adults. (Poindexter, 2008a, p. 18)

What's tricky in the socialization process that may contribute to the development of a news engagement gender gap is that boys and girls may be socialized differently when it comes to news. In fact, a journalism scholar studying parents' socialization influence on their children's newspaper reading observed: "It usually is expected in our society that boys will become concerned about public affairs.… Girls usually are not expected to become absorbed with political developments" (Clarke, 1965, p. 545).

Furthermore, one of the advantages of a mobile device—small enough to carry in your hand or pocket, which can make getting news more convenient—is actually a disadvantage when it comes to socialization by observational learning. Through observational learning, children can learn to engage with news by observing their parents engaging with news, whether reading a newspaper or watching local or network news. But on a mobile device, unless parents explicitly tell their children that they are engaging with news, parents could be posting on social media, searching for information, playing a game, or even banking. That's why as a socialization agent for developing future news consumers, mobile devices are expected to be less influential, but it will take another generation—the post-Millennial Generation—before we know for sure. Millennials are the children of Baby Boomers and Generation X, and as they were coming of age, their parents were reading print newspapers, starting to read news on laptops, and watching local and network news on TV. Now that the first Millennials are starting their own families, they will be the role models for their children in accessing news, but much of their news engagement will take place on mobile devices, which means the indirect socialization influence may be muted.

How a generation socializes the next generation about the importance of being informed may be related to how that generation was raised. That's why survey participants were asked to strongly agree or strongly disagree with the following statement on a five-point scale: "You were raised to believe being informed of the news is important." A smaller percentage of Millennials than Baby Boomers and Generation X strongly agreed with the statement. Baby Boomers and Generation X strongly agreed at about the same level (36% vs. 37%), but there was a gender gap among Millennials, with males less likely than females to strongly agree with the statement (23% vs. 30%). These results suggest that this belief is not being passed on to the next generation at the same level as the previous generation but, perhaps, more important, these results reveal a broader trend that is far removed from the 1964 institutional behavior of being informed by reading a newspaper. The overwhelming majority of all generations do *not* strongly endorse the statement they were raised to believe being informed of the news is important.

If one is raised to believe that being informed of news is important, does that translate into believing being informed is a civic responsibility similar to voting? The civic responsibility to be informed, which is "learned in the formative years" through socialization was found to exist in a pre-Internet world, but its strength weakened somewhat in an Internet world (McCombs & Poindexter, 1983; Poindexter & McCombs, 2001). The question is: In a mobile world, do people believe they have a civic responsibility to be informed? To answer that question, survey participants were asked to strongly disagree or strongly agree on a five-point scale with the statement, "We all have a duty to keep ourselves informed about news and current events." Males and females in the older Baby Boomer and Generation X generations equally agreed that we had a responsibility to be informed of news and current events (38% vs. 38%), but among Millennials, males were less likely than females to strongly agree that we have a civic responsibility to be informed (24% vs. 35%).

Table 7.9 shows how these beliefs, which are the result of the socialization process, can translate into news seeking in a mobile world. For example, the first two columns compare those who were raised to believe being informed is important (strongly agree) versus those who were not raised to believe being informed is important (strongly disagree). For males and females of every generation, being raised to believe being informed is important matters in seeking news daily. Specifically, there is a percentage point difference of 26 to 35 points in terms of seeking news daily. Table 7.9 also reveals both a generational difference and a gender differential with being raised to believe being informed is important. In other words, being raised to believe it's important to be informed correlates with daily news seeking more for Baby Boomer and Generation X males and females than Millennial males and females. But within generations, males who were socialized to believe being informed is important were more likely than females to seek news every day: 75% of Baby Boomer and Generation X males compared to 57% of their female counterparts sought news daily and 43% of Millennial males compared to 33% of Millennial females sought news daily.

The gender divide shrinks in older and younger generations when it comes to believing there is a responsibility to be informed and daily news seeking; 73% of older males compared to 60% of older females who believe there is a responsibility to be informed seek news daily, while 40% of Millennial males, compared to 36% of Millennial females, who strongly believe there is a responsibility to be informed seek news every day.

Table 7.9: Comparing Socialization and Responsibility to Be Informed and Seeking News by Gender and Generation.
% Seeking News Daily

	Strongly Disagree "Raised to believe being informed of the news is important"	Strongly Agree "Raised to believe being informed of the news is important"	Strongly Disagree "We have a duty to keep ourselves informed about news and current events."	Strongly Agree "We have a duty to keep ourselves informed about news and current events."
Baby Boomer/ Generation X males	40%	75%	54%	73%
Baby Boomer/ Generation X females	25%	57%	11%	60%
Millennial males	17%	43%	17%	40%
Millennial females	7%	33%	0%	36%

Being taught it is important to be informed while growing up is vital in the development of a news-seeking adult. But if the news fails to deliver what consumers expect, news may become a low priority even for individuals who feel they have a responsibility to be informed. In addition to what the news is expected to deliver, engaging with news and becoming informed can produce unexpected benefits that have the potential to strengthen the relationship between consumers and news. Enjoyment, empowerment, and helping with daily life are not benefits that people first think of when talking about engaging with news but, as Table 7.10 shows, they can make a difference in daily news seeking.

Not surprisingly, men and women of older and younger generations who strongly enjoy keeping up with news are significantly more likely than those who don't enjoy keeping up with news to seek news daily. And the generational difference was also not a surprise: Baby Boomer and Generation X males and females who strongly enjoyed keeping up with news are significantly more likely than Millennial males and females to seek news every day. But what is somewhat of a surprise is the gender gap among older generations who strongly enjoy keeping up with news but not the younger

Millennials. Baby Boomer and Generation X males who strongly enjoy keeping up with news are significantly more likely than their female counterparts to seek news every day (74% vs 66%) while Millennial males and females who strongly enjoy keeping up with news seek news every day at the same level (47% vs. 47%).

Survey participants were asked to strongly agree or strongly disagree with a statement that tried to capture a sense of empowerment that people may feel as a result of being informed of the news. The empowering effect of news can make a difference in seeking news daily; people who feel empowered by being informed are significantly more likely than those who don't feel empowered to seek news every day. But this sense of empowerment from being informed also reveals a gender gap where both Baby Boomer and Generation X females and Millennial females lag behind their male counterparts in seeking news daily. Almost three-quarters (74%) of Baby Boomer and Generation X males who strongly agree news makes them feel empowered seek news every day compared to 63% of their female counterparts. Similarly, 50% of Millennial males who strongly feel news is empowering seek news daily compared to 41% of Millennial females.

Overall, survey participants who strongly agreed that news helped with daily life were significantly more likely to seek news every day, but there were gender differences in the older and younger generations. Baby Boomer and Generation X males (78%) who strongly agreed that news helped with daily life were significantly more likely than their female counterparts (65%) to seek news daily. Similarly, Millennial males who strongly agreed that news helped with daily life were more likely than Millennial females to seek news daily (63% vs. 45%).

Table 7.10: Benefits of Being Informed and Daily News Seeking by Gender and Generation. % Seeking News Daily

	Enjoy (Strongly Agree)	Empower (Strongly Agree)	Help with Daily Life (Strongly Agree)
Baby Boomer/ Generation X males	74%	74%	78%
Baby Boomer/ Generation X females	66%	63%	65%
Millennial males	47%	50%	63%
Millennial females	47%	41%	45%

Public's Expectations of the News Media: Watchdog, Good Neighbor, Free

Socialization and benefits received from being informed influence news engagement but so do the public's expectations, evaluations, and trust of the news media. Four decades of national surveys of journalists beginning in the 1970s had found the majority of journalists viewed their role as a "watchdog" but it wasn't until the pre-mobile world of 2006 that the public was asked about journalists' watchdog role and other expectations of the press (Willnat & Weaver, 2014; Poindexter et al., 2006). Results found that some endorsed the watchdog role but others preferred the press, especially local news, to be more like a good neighbor (Poindexter et al., 2006). To determine if the watchdog and good neighbor expectation are also present in today's mobile landscape, survey participants were asked to strongly agree or strongly disagree on a five-point scale with the following two statements: "It's important for the news media to be a watchdog of powerful people and the government" and "The news media should be more helpful and caring like a good neighbor." An analysis found that both watchdog and good neighbor are correlated with seeking news every day.

Specifically, the expectation that the press should be a watchdog had a positive effect on daily news seeking for Baby Boomer and Generation X females as well as Millennial males and females. Fifty-six percent of Baby Boomer and Generation X females who strongly endorsed the watchdog role for the press sought news every day compared to 29% who strongly disagreed with a watchdog role for the news media. Similarly, more Millennial males who strongly endorsed a watchdog role compared to Millennial males who strongly disagreed with a watchdog role for the press sought news every day (51% vs. 33%). Millennial females who strongly agreed with a watchdog role were more likely than Millennial females who strongly disagreed with a press watchdog role to seek news on a daily basis (46% vs. 0%). Oddly, the relationship between endorsing a watchdog press role and daily news seeking was the reverse for Baby Boomer and Generation X males. Nine out of ten (88%) of Baby Boomer and Generation X males who strongly disagreed with a watchdog role for the press sought news every day compared to 64% who strongly endorsed a watchdog role.

The relationship between the good neighbor press role and daily news seeking was only relevant for the Millennial generation. Over half (53%) of Millennial males who strongly agreed "the news media should be more helpful and caring like a good neighbor" sought news every day compared to 33% who strongly disagreed with the good neighbor role for the press. The good neighbor expectation for Millennial females didn't drive daily news seeking as much as it did for Millennial males, yet 44% of Millennial females who strongly agreed the press should be more like a good neighbor sought news daily compared to 33% of Millennial females who strongly disagreed with the concept of the press as a good neighbor.

While the watchdog and good neighbor roles are more directly related to how the public expects the press to function in society, the public also has an expectation about whether news should be free on the Internet. Survey participants were asked to strongly agree or strongly disagree on a five-point scale with the statement, "All news on the Internet should be free to the public," and males and females of both older and younger generations were compared to see if the expectation of free news on the Internet was a factor in seeking news every day. Baby Boomer and Generation X males and females, as well as Millennial males, who strongly agreed that news on the Internet should be free were more likely to seek news every day than those who strongly disagreed that news on the Internet should be free. The following are specific percentages in seeking news daily for those who strongly agreed versus those who strongly disagreed news should be free on the Internet: Baby Boomer and Generation X males (61% vs. 42%); Baby Boomer and Generation X females (48% vs. 33%); Millennial males (35% vs. 0%). Millennial females had a curvilinar-shaped pattern of daily news seeking: Millennial females who strongly agreed news should be free on the Internet as well as those who strongly disagreed, about equally sought news every day (31% vs. 33%) while those with more moderate opinions about free news on the Internet sought news less frequently.

Evaluating Press Performance

The public's expectations of the press are not limited to its watchdog or good neighbor role or even if news should be free on the Internet. Many of the expectations the public has of the press can be viewed in the context of fair, balanced, inclusive, and accurate coverage and ethical journalistic principles and processes. So by evaluating the press's performance in these areas, additional insight can be gained about what the public expects of the press.

On one hand there's no reason to expect evaluations of the press's performance should differ in a mobile news landscape. But on the other hand, because of the speeded-up pace of reporting and updating news in a 24/7 mobile-social media world, as well as the availability of a mobile device to check the news anytime and from anywhere, the potential of inaccuracy and awareness of that inaccuracy are magnified.

Survey participants were therefore asked to evaluate press performance in the following six areas on a scale of 1 to 5, with 1 being strongly disagree and 5 being strongly agree: prioritizing being first over being right, willingness to admit mistakes, accurate representation of constituent groups, context, bias, and accuracy. Of the six areas, bias and accuracy were unrelated to daily news seeking in the context of gender and generation. For the other four areas, some were related to seeking news daily in the context of their gender and generation and others were not.

Baby Boomer and Generation X males who strongly *disagreed* with the statement, "News organizations are more concerned about being first than being right" were more likely to seek news seven days than Baby Boomer and Generation X males who strongly *agreed* with the statement (80% vs. 62%). Similarly, Millennial males who strongly *disagreed* that news organizations were more concerned about being first than right sought news seven days a week compared to Millennial males who strongly *agreed* with the statement (100% vs. 37%).

While older and younger males rejected the idea that news organizations prefer being first over being right, females from the older generations rejected the assertion, "The news does a poor job of providing the context to understand a story." Baby Boomer and Generation X females who strongly *disagreed* with that statement were more likely to seek news seven days a week than females who strongly *agreed* with the statement (58% vs. 43%). For these same women, the news media's willingness to admit mistakes made a difference in seeking news every day; Baby Boomer and Generation X females who strongly agreed that "news organizations are willing to admit their mistakes" were more likely to seek news every day than females in their same generational cohort who strongly disagreed with the statement (52% vs. 38%).

The statement, "The news projects an accurate representation of the constituent groups in U.S. society" made a difference in seeking news seven days a week for both Millennial males and females but not for their counterparts among older Baby Boomers and Generation X. Almost three-fifths (59%) of Millennial males who strongly agreed with the statement sought news seven days a week compared to 9% of Millennial males who strongly disagreed with the statement. For Millennial females, 34% who strongly agreed that the news projects an accurate representation of the constituent groups in U.S. society sought news every day compared to 11% who strongly disagreed with the statement.

The Trust Factor

If the quality of democratic life in the U.S. depends on the public having the facts and being able to make sense of them, as the authors of *The Elements of Journalism* assert in the most recent edition of their book, then both the source of the facts and trust of the sources of the facts are essential in a democracy. While journalists and news organizations are not the only sources of the facts required for a democratic society to function, when serving as a watchdog, journalists and news organizations may be the only independent sources of facts. Understanding the level of trust the public has for the news media and journalists is essential because trust can facilitate or inhibit the public's learning of facts necessary for a functioning democracy. Additionally, trust has the potential to be a factor in news engagement in general and engaging with news distributed by various news sources. If the public doesn't trust the messenger, would the public turn to them to get the news?

Do males and females trust the press the same or is there a gender gap similar to gender gaps identified in news engagement, socialization, expectations, and performance evaluations? Table 7.11 finds some gender and generational differences in trust and reminds us that the press is not a monolith. Trust varies for local and national news as institutions and journalists who are generally connected to news media institutions but are evaluated as an entity that transcends institutions. Additionally, social media, which have emerged in the news media landscape as an indispensable news delivery and sharing platform, also have trust issues in the context of gender and generation.

Survey participants were asked whether they trusted the sources for news listed in Table 7.11 "a lot," "some" or "none." The results found local sources of news were trusted more than national sources of news, and journalists and social media were the least trusted sources of news.

A gender gap in trust surfaced in local news but not national news or among journalists. For both older Baby Boomers and Generation X, as well as younger Millennials, more females than males trust local TV news (47%

Table 7.11: Trust of Local News, National News, Journalists, and Social Media by Gender and Generation.
% Saying Trust A lot

	Baby Boomer & Generation X Males	Baby Boomer & Generation X Females	Millennial Males	Millennial Females
Local News:				
Local TV news	42%	47%	36%	40%
Local print and online newspaper	33%	37%	26%	35%
National News:				
Network TV*	28%	32%	27%	26%
Cable news	26%	25%	29%	30%
National newspaper*	24%	26%	26%	32%
Journalists	15%	18%	18%	19%
Social media:				
Facebook	10%	15%	22%	18%
Twitter	7%	7%	20%	13%

*Differences are not statistically significant.

vs. 42% and 40% vs. 36%) and local newspapers, including online and print (37% vs. 33% and 35% vs. 26%). Despite the overall low trust level for social media, there was a generational and gender difference in trust. Not surprisingly, Millennials more than older generations trust social media for news, and more Millennial males than Millennial females trust Twitter as a news source (20% vs. 13%). For Baby Boomers and Generation X, Facebook is more trusted as a news source for females than males (15% vs. 10%).

Is the News Consumption Gender Gap Now the Norm?

Almost everywhere you turn in these data about the mobile news landscape, a gender gap is lurking, suggesting that the pre-mobile gender gap documented in *Women, Men, and News: Divided and Disconnected in the News Media Landscape* did not disappear but may have gathered steam and become institutionalized (Poindexter et al., 2008). An institutionalized gender gap is not good for democracy, and it's not the way to achieve a society where males and females are equally informed about the facts they need to fully function in a democratic society. The reasons for the gender gap with females getting and seeking news less than males may be due to the fact that social media are competing for consumers' attention and time and are more successful with women; a failure for the benefits received from being informed to resonate equally among males and females; and the continued failure of news coverage to equally include males and females. Unfortunately, these reasons may be exacerbated in the mobile space because social media are easily accessible on always-available smartphones, and when news is in direct competition with social media for the time and attention of consumers, news usually loses. But the news consumption gender gap does not have to be the norm in the mobile landscape, that is, if the recommendations proposed in Chapter 10 are implemented.

· 8 ·

POLITICAL IDENTITY, NEWS, SOCIAL MEDIA, AND MOBILE

The 2008 presidential election in which a junior senator from Illinois beat the senior senator from Arizona was historic. For the first time in the history of the United States, an African American was elected president. Less well known and, for many, significantly less important, the 2008 presidential election also made history because the Obama campaign led the presidential election process out of what his campaign manager called the technological "Dark Ages" (Plouffe, 2009, p. 36). By moving from the Dark Ages of technology to digital enlightenment, the Obama campaign connected with Millennials, a must-win new voter if then-senator Barack Obama had any chance of winning the White House, and established a new standard for presidential campaigns in the digital age. The Obama campaign's commitment to digital, as well as reaching Millennials, was evident in its first campaign video released on the upstart digital video-sharing site YouTube that Millennials adored and older generations had likely never heard of.

The Obama strategy to target Millennials paid off because without the Millennial and African American vote the 2008 election may have had a different outcome. Digitally targeting Millennials not only turned out to be a winning strategy but it also forever changed how presidential campaigns would be run. By the 2012 presidential campaign, both President Obama and Republican presidential candidate former Massachusetts governor Mitt Romney targeted Millennials and older voters using digital platforms whether through YouTube or social media, especially Twitter, which was a must-report platform for political journalists as well as a must-follow platform for Millennial voters who wanted to keep up with the candidates and issues (In swing states, 2012).

By the 2012 presidential election, the iPhone had been around five years and smartphone adoption was increasing rapidly. The iPad was introduced two years before the 2012 presidential election, so it was, perhaps, more a media darling than an adoption sensation. Smart mobile device ownership was growing but not enough to transform the 2012 presidential election as it has been doing during the 2016 race to the White House. In Iowa, New Hampshire, Nevada, and South Carolina, where Democrats and Republicans held their first caucuses and primaries, voters were not just getting up close and personal, they were pulling out their smartphones and taking selfies with candidates. And because smartphones were voters' 24/7 companions, anytime and from anywhere, voters could follow their preferred candidates, click on election trending topics or tap an app to access the latest news about candidates, issues, debates, and horserace polls from Fox, CNN, MSNBC, the *New York Times*, the TV networks, and other news sources.

Candidates also asked voters to pull out their phones and text them their phone numbers. In 2016, smartphone numbers replaced email, which had replaced snail mail, as the best way to connect with voters because texting enabled campaigns to "get a message out as widely and quickly as possible, and with confidence that it will be read" (Corasaniti, 2015). Plus, because texting is how Millennials communicate, texting allows campaigns to easily and instantaneously reach the largest generation of prospective voters in the 2016 presidential election (Corasaniti, 2015).

What will be the lasting impact of the ubiquitous smartphone on the 2016 presidential election? Will voters be more informed because news about the election is always accessible with just a finger tap? Will voters feel more engaged and empowered, which could lead to feeling that their vote counts? Will smartphones matter in voter turnout, which ultimately can be the difference maker in who wins and who loses a presidential election?

We won't know the turnout until after the 2016 presidential election but it will need to be higher than the 61.8% turnout in 2012, which was actually down from 63.6% in the historic election of 2008 when the first African American president was elected (The diversifying electorate, 2013).

Furthermore, African Americans, who for the first time in history voted at a higher rate than whites, as well as Millennials may be the difference makers and so will Independents who can go either way. Finally, if the gender gap that has been a factor in U.S. presidential elections does not close, it may be easy to predict the outcome of the 2016 presidential election. But because Republican and Democratic primary voters have not followed the script of past elections, it's virtually impossible to predict what will happen November 8, 2016, when Americans elect the 45th president of the United States.

The ubiquity of smartphones may be playing a role in the unpredictability on the 2016 presidential campaign trail because voters have their own independent access to candidates and their stands on issues. Voters can use their smartphones to read media reports, watch live streaming campaign events, or follow candidates with their Twitter, Facebook, and Instagram apps. But whether it's due to unfiltered access to candidates through mobile devices, anger and distrust of traditional candidates, income inequality and stagnant wages, fear of terrorism, or lopsided news coverage that spotlights a billionaire celebrity political novice who has uttered critical and offensive statements about undocumented immigrants, Latinos, women, Muslims, the disabled, journalists, war heroes, past presidents of their own party, and even the Pope, the 2016 presidential race will likely be one for the history books for both its head-scratching unpredictability and the impact of mobile devices.

As the first U.S. presidential election to be contested in a mobile landscape it's too early to know if differences will emerge in the ways Democrats, Republicans, Independents, and even Tea Party members use their mobile devices to follow candidates and issues. But there is some evidence from the News for a Mobile-First Consumer Survey that mobile news activity is similar across political parties.

Demographic and Ideological Variations across Political Parties

The political differences across Democrats, Republicans, and Independents do not necessarily translate to some of the mobile, news, and social media differences found in earlier chapters, but they do translate to demographic distinctions. For example, Democrats are more likely to be female and Republicans are more likely to be male. Democrats have more African Americans and Hispanics and Republicans are more likely to be white. Wave I Millennials are more likely to be found among Democrats while Wave II Millennials are more likely to be found among Independents. Although there are no statistically significant educational differences among Democrats, Republicans, and Independents, there are income differences: more Democrats and Independents have an annual household income of under $40,000, while more Republicans have a household income above $80,000.

Not surprisingly, ideology is also related to political party. Liberals or progressives are more likely to be found among Democrats while conservatives and Tea Party members are more likely to be found among Republicans. Survey participants who identify as "middle of the road" are more likely to be found among Independents.

Studies have found that one's ideological stance can influence news that individuals expose themselves to. Specifically, news that is purposely produced to have an ideological bent has been found to appeal to those who hold similar ideological beliefs (Mitchell, et al., 2014). But are there political party differences in getting and seeking news in general and news about topics that are directly or indirectly related to election issues or topics that may drive the election? And do different political groups—Democrats, Republicans, and Independents—have preferences for certain news platforms?

While there is no significant difference across political parties in seeking news, there is a difference in getting news. According to Table 8.1, more Republicans (63%) get news seven days a week than Democrats (59%) and Independents (54%). Plus, for some types of news often paid attention to, differences are related to political party.

To provide further insight into the news that Democrats, Republicans, and Independents get and seek, they were compared on eleven news topics that should matter to these three political groups. Some of the news topics like politics and elections are more directly related to political party, and others, while indirectly related to political party, are topics that informed and engaged voters should have knowledge about. Interestingly, for seven of the eleven topics, there was no statistically significant difference in paying attention to the topic often; in other words, the three political groups were similar in paying attention to these topics often. The topics that Democrats, Republicans, and Independents differed in paying attention to often were election, government, health, and environment news topics that often take center stage during elections, partisan debates, and controversies.

Table 8.1: News Engagement and Political Party.

	Democrats	Republicans	Independents
Get news 7 days per week	59%	63%	54%
Seek news 7 days per week*	45%	47%	42%
Often pay attention to international news*	42%	44%	41%
Often pay attention to national news*	58%	62%	56%
Often pay attention to local news*	72%	72%	66%
Often pay attention to health news	26%	20%	20%
Often pay attention to business news*	18%	23%	18%
Often pay attention to politics*	31%	36%	28%
Often pay attention to government	31%	40%	30%
Often pay attention to natural disaster news*	39%	38%	37%
Often pay attention to election news	27%	32%	24%
Often pay attention to crime news*	35%	34%	33%
Often pay attention to environment news	22%	16%	21%

*Differences are not statistically significant.

Democrats, Republicans, and Independents Connecting in the Mobile Landscape

Democrats, Republicans, and Independents may differ ideologically but they are similar in their relationship to their mobile devices. Approximately two-thirds of the three political groups are mobile laggards; about one-third are mobile specialists, and a small percentage are mobile-first, consumers who use their mobile devices for everything. Whether mobile-first, mobile specialists, or mobile laggards, there is no significant difference among Democrats, Republicans, and Independents in the platforms used to access news.

Table 8.2: Platforms Used to Access News by Political Party.*

	Democrats	Republicans	Independents
TV	40%	40%	42%
Mobile device	23%	17%	21%
Laptop	22%	24%	20%
Print paper	7%	10%	8%
Other	5%	6%	6%
Radio	2%	4%	3%

*Differences are not statistically significant.

In fact, Democrats, Republicans, and Independents are not only similar in the platforms used to access news, they are similar in relying on TV more than any other platform to access news (see Table 8.2). Additionally, in just over a half decade, mobile devices are challenging laptops as the second most popular platform to access news for Democrats, Republicans, and Independents. This is a significant change from the pre-digital age when print newspapers were the second most popular place to get news. Today, 10% or fewer Democrats, Republicans, and Independents turn to print newspapers for news.

The popularity of mobile devices for accessing news can also be gleaned from the percentage of time used on a mobile device to get news. Democrats are significantly more likely than Republicans and Independents to spend at least 20% of mobile device time on news. More than three-fifths (63%) of Democrats compared to 48% of Independents and 43% of Republicans spend at least 20% of their time on a mobile device accessing news. An even larger percentage of Tea Party members spend at least 20% of mobile time engaging with news; 75% of Tea Party members compared to 51% of non-Tea Party members spend at least 20% of mobile time engaging with news.

The differences found in amount of time using a mobile device for news does not translate into differences in using news apps. Just as there is no statistical difference in using mobile devices for news, there is no difference among Democrats, Republicans, and Independents in using apps or not using apps to get news. Still Democrats, Republicans, and Independents are more likely to not use news apps than to use them. But for those who do use news apps, 30% of Democrats have three or more news apps compared to 27% of Republicans, and 22% of Independents.

Despite the fact that Democrats, Republicans, and Independents are more likely to be non-news-app users than users, apps are the most popular gateway to news on mobile devices, followed by browsers, Facebook, and breaking news alerts. The least popular methods for accessing news on mobile devices are Twitter, search engines, and notifications. Regardless of method used to access news on smartphones and tablets, any differences among Democrats, Republicans, and Independents are not statistically significant. But Democrats, Republicans, and Independents do differ significantly in engaging with social media, including downloading social media apps on their mobile devices and engaging with news through Facebook and Twitter.

Engaging Social Media

Just as Facebook dominates in the world of social media, Facebook dominates in the world of political parties and Independents. More Democrats, Republicans, and Independents are on Facebook than other social media and more of them have a Facebook app on their mobile device than other social media apps. But there is one place Facebook does not dominate: devoting at least one-third of social media time to news. A slightly larger percentage of Democrats, Republicans, and Independents spend at least one-third of Twitter time on news compared to the political groups that spend time on Facebook accessing news.

Twitter may not be the most popular gateway to news on mobile devices but Table 8.3 underscores that for Democrats, Twitter provides access to news in several important ways. Democrats are more likely than Republicans and Independents to access news on Twitter by following journalists, clicking tweeted news, and retweeting news. Connecting to news through Twitter, whether on a mobile device or not, probably explains why Democrats trust Twitter for news more than Republicans and Independents. Still, when Facebook and Twitter are compared head-to-head on trust, slightly more Democrats, Republicans, and Independents trust Facebook over Twitter as a news source.

Table 8.3: Comparing Democrats, Republicans, and Independents on Social Media, Apps, and News Engagement.

	Democrats	Republicans	Independents
Facebook:			
On Facebook	80%	74%	72%
Have Facebook app	66%	58%	58%
Spend at least 1/3 Facebook time on news	30%	20%	20%
Trust Facebook as source of news a lot	16%	13%	12%
Twitter:			
On Twitter	40%	31%	32%
Have Twitter app	36%	28%	29%
Spend at least 1/3 Twitter time on news*	35%	27%	23%
Follow journalist on Twitter	48%	40%	32%
Click tweeted news often	29%	21%	11%
Retweet news often	18%	13%	6%
Click trending news often*	26%	22%	13%
Trust Twitter as source of news a lot	12%	8%	6%
YouTube and Instagram:			
Have YouTube app	62%	50%	56%
On Instagram	31%	22%	24%
Have Instagram app*	33%	22%	27%

*Differences are not statistically significant.

More Democrats than Republicans and Independents also have apps for YouTube and Instagram, and when apps for the video-sharing YouTube are compared with Instagram, which allows for sharing of photos and videos, YouTube apps win. At least twice as many Democrats, Republicans, and Independents have YouTube apps than Instagram apps. The YouTube app underscores how much has changed in just eight years. When the 2008 Obama campaign released its first video on YouTube, it could only have been viewed from a laptop or desktop. But today in the mobile landscape, a presidential candidate's video can be viewed with a finger tap on an app while eating in a restaurant, exercising on a stationary bike in the gym, sitting in a classroom, riding as a passenger in a car, sunbathing on a beach, or picnicking in the park.

The Mobile Transformation of the Presidential Election Landscape

As the 21st century dawned and more and more of the public adopted the Internet to communicate and access news and more news organizations produced news digitally, the legacy media landscape of the 20th century gave way to a 21st century digital media terrain that would become populated with social media, providing innovative ways to connect, communicate, and facilitate the distribution of and engagement with news.

As smartphones emerged and were adopted, they redefined everything from how consumers communicated and engaged with social media and news to how consumers played games, listened to music, and watched TV and movies, to how they banked, shopped, and carried out basic daily living tasks from setting an alarm clock to getting directions. And for the first time smartphones redefined how voters experienced a presidential election. Prospective voters could watch and read news about the candidates and even communicate with presidential hopefuls on the campaign trail anytime and from anywhere using their smartphones.

As the first mobile presidential election, 2016 will have forever changed how candidates campaign and how voters keep up with them. The immediacy and convenience of access to candidates and news about them that mobile devices provide can change the dynamics of a presidential race. For example, campaigns have to become more vigilant than ever in responding to press criticisms and attacks by rivals. In the mobile landscape, presidential candidates no longer have a 24-hour news cycle; in the mobile landscape, the response window may be as short as the time it takes to tweet.

Just as smart mobile devices created a new type of consumer, the 2016 presidential election is seeing the emergence of a new type of voter or at the very least traditional voter constituencies affected in new ways. While we won't know the full effect of mobile until November 2016, results from the Survey for a Mobile-First Consumer provide insight into four constituencies that will be factors in winning the White House in 2016: Independents, women, Millennials, and voters of color.

The Independent Voter

The Independent voter presumably requires lots of information to make a decision about which candidate to vote for because the Independent is not anchored to a specific political party. In fact, that's what makes the Independent voter so powerful in an election; she or he can be wooed by candidates on both sides. Of course, there's always the possibility that the Independent voter will skip the election and not vote. While the Independent voter did not skip the election in 2012, the Independent voter did not vote at the same rate as Democrats and Republicans, according to the Roper Center.

Of those who turned out to vote in 2012, 29% were Independents, 38% were Democrats, and 32% were Republicans, according to a 2013 Roper Center analysis. While not surprisingly the vast majority of Democrats voted for the Democratic presidential candidate and an overwhelming majority of Republicans voted for the Republican presidential nominee, Independents split their vote, with 45% supporting President Obama and 50% voting for former Massachusetts governor Mitt Romney (How groups voted, n.d.).

Voting is an important dimension of civic engagement, but it is, of course, not the only way citizens participate in the civic arena. Voting, though, may be one of only a few political or civic activities in which citizens traditionally have felt a duty to participate. Studies have found a similar duty to keep informed (McCombs & Poindexter, 1983; Poindexter & McCombs, 2001). This duty to be informed has contributed to higher levels of news engagement.

Fewer Independents voting in the 2012 presidential election suggests Independents may feel less duty bound to vote, but according to Table 8.4, Independents do not feel less duty bound to keep informed. Although the overwhelming majority of Independents, Democrats, and Republicans do not feel a duty to get informed, there is no statistically significant difference among them. But there is a difference in paying attention to public affairs news, an index that combines news about politics, government, and elections. Republicans (42%) are significantly more likely than Democrats (34%) and Independents (32%) to pay frequent attention to public affairs news. Ironically, paying more attention to public affairs news does not make Republicans feel more empowered by being informed. In fact, Democrats feel more empowered, followed by Republicans. Independents feel the least empowered from being informed.

While contacting an elected official and donating money to a campaign are civic activities to participate in, they are not necessarily tied to a civic duty. Still, Republicans are significantly more likely than Democrats to make a campaign contribution, but in terms of contacting an elected official, the three political groups do not differ statistically.

Table 8.4: How Independents Compare with Democrats and Republicans in Civic and News Engagement and Attitudes.

	Democrats	Republicans	Independents
Donated money to campaign or political cause	22%	16%	10%
Contacted elected official*	26%	25%	20%
Duty to keep informed*	40%	36%	34%
Frequent attention to public affairs news	34%	42%	32%
Feel empowered from being informed	23%	18%	16%
Watchdog press*	30%	28%	32%
Help protect democracy	20%	8%	11%

*Differences are not statistically significant.

Do Independents view the press's role the same as those who are members of an established political party, especially when the role is closely aligned with the process of journalism in covering public affairs, elections, and government? When asked to strongly disagree to strongly agree with the statement, "It's important for the news media to be a watchdog of powerful people and the government," there's no statistical difference among the three groups. In other words, Independents are similar to Democrats and Republicans in strongly endorsing the concept of the press as watchdog. While all groups equally endorse the watchdog role for the press, they do not share the same view about whether the press protects democracy or harms it. When asked to strongly agree or strongly disagree, only one-fifth or fewer strongly agreed that "news organizations protect democracy more than hurt democracy," with Democrats (20%) twice as likely as Republicans (8%) and Independents (11%) to strongly endorse the idea of the press as a protector of democracy.

Tea Party in the Political Landscape

While Democrats, Republicans, and Independents differ in the majority of political, civic, and news-related activities and perceptions, it is important to underscore that some of the differences may be greater because within the two political parties and Independents, there are conservatives, moderates, and liberals or progressives. In fact, Republicans are three times more likely to be conservative than Democrats and Independents. And Independents are significantly more likely than Democrats and Republicans to be moderates, while Democrats are far more likely than Republicans and Independents to be liberals. Finally, Tea Party members are primarily a part of the Republican Party but some can be found among Democrats and Independents. Tea Party members have been described as a "mixture of libertarian, populist, and conservative activists" (Tea Party movement, n.d.). About 10% of Americans identify with the Tea Party, which differs from non–Tea Party members in some political and civic activities and perceptions of the press.

Tea Party members are more likely than non–Tea Party members to have donated money to a campaign (43% vs. 13%) and contacted an elected official (50% vs. 21%). Tea Party members are significantly more likely than non–Tea Party members to strongly agree that they have a duty to keep informed (49% vs. 35%), and they are almost twice as likely to pay frequent attention to public affairs news (60% vs. 32%) and feel empowered from becoming informed (27% vs. 18%). A larger percentage of Tea Party members than non–Tea Party members spend at least 20% of their mobile device time on news (75% vs. 51%). Tea Party members also strongly endorse the watchdog role of the press (48% vs. 28%) and Tea Party members strongly agree that news outlets protect democracy more than hurt it (24% vs. 13%).

Gender Gaps in the Political Arena Landscape

Just as gender gaps in society, the workplace, STEM fields, and newsrooms are not news, gender gaps in the political arena are also not news. What seems to become news every four years is how large or small the gender gap will be. In the 2012 presidential election, more women than men voted (53% vs. 47%), and there was an 11-point gender gap between the Democratic and Republican presidential candidates; 55% of women voted for President Obama, the

Democratic presidential candidate, and 44% voted for the Republican presidential candidate Mitt Romney (How groups voted, n.d.).

Gender gaps are not just about differences across political parties. As Table 8.5 shows, gender gaps can be within political parties, especially when it comes to political behavior such as contacting an elected official and contributing to a campaign. Across Democrats, Republicans, and Independents, gender gaps range from four to eleven percentage points, with more males than females contacting elected officials or contributing to a campaign. Paying frequent attention to public affairs news produces an even larger gender gap: 12 to 19 percentage points. Democratic, Republican, and Independent men pay significantly more attention to public affairs news than their female counterparts.

Table 8.5: Gender Gaps in the Political, Civic, and News Landscape.

	Male Democrats	Female Democrats	Male Republicans	Female Republicans	Male Independents	Female Independents
Contacted elected official	30%	23%	30%	19%	24%	17%
Donated money to campaign or contributed to political cause	27%	18%	21%	10%	12%	8%
Frequent attention to public affairs news	45%	26%	47%	35%	40%	23%
Duty to keep informed*	38%	42%	34%	38%	35%	34%
Feel empowered from being informed	21%	25%	17%	20%	16%	17%
Watchdog press	35%	26%	32%	23%	36%	28%
Protect democracy	22%	17%	7%	9%	13%	9%

*Differences are not statistically significant.

And although there is a gender gap in frequently paying attention to public affairs news across Democrats, Republicans, and Independents, there is no statistically significant gender gap in the belief that there is a duty to keep informed. And despite paying less attention to public affairs news than their male counterparts, Democratic and Republican women feel slightly more empowered by being informed than the men in their respective political parties.

Because the press's role as a watchdog and a protector of democracy have political and civic implications, they were analyzed to determine if men and women of the different political parties were in agreement and the results showed similarities and differences. There was a gender gap in each political party and among Independents. Males were more likely than females for Democrats (35% vs. 26%), Republicans (32% vs. 23%), and Independents (36% vs. 28%) to very strongly endorse the watchdog role of the press.

And while fewer than a quarter of the two political parties and Independents strongly agreed with the statement, "News organizations protect democracy more than hurt democracy," the gender gap was among Democrats and Independents. Democratic men were more likely than women (22% to 17%) and Independent men were more likely than women (13% vs. 9%) to strongly agree that news organizations protect democracy more than hurt it. Among Republicans, only 7% of males and 9% of females strongly agreed that the press protects more than harms democracy.

Wave I and Wave II Millennials in the Political Arena

In 2012, Millennials born between 1995 and 1999 were too young to vote so they were not among the 19% of Millennials who voted during that year's presidential election. Of the Millennials who voted, 60% supported President Obama and 37% supported former Massachusetts governor Mitt Romney. While more voters in the 30 to 44-year-old age group supported President Obama than the Republican candidate (52% vs. 45%), more of the voters who were 45 or older supported Mitt Romney over Barack Obama (How groups voted, n.d.). Specifically, 51% of voters supported Romney compared to 47% who supported Barack Obama. Fifty-six percent of 65-plus supported Romney compared to 44% who supported Barack Obama.

Since all Millennials, except those born in 1999, who register to vote will be eligible to vote in the general election for president in 2016, the question is: Will younger Wave II Millennial voters follow in the political, civic, and news engagement footsteps of older Wave I Millennials? While the answer to that question won't be known until November 2016, there are indicators in the Survey of Mobile-First Consumers that Wave II Millennial Democrats, Republicans, and Independents are both similar to and different from their older Millennial counterparts.

There are no statistical differences among Wave I and Wave II Millennial Democrats, Republicans, and Independents in three areas that have implications for political news engagement and the duty and consequences of being informed: paying attention to public affairs news frequently, belief in a duty to keep informed, and feeling empowered by being informed (See Table 8.6.).

But there are differences in activities related to contacting elected officials and contributing to campaigns. Wave I Millennials who are Democrats and Independents are more likely than their Wave II counterparts to have contacted an elected official, but among Republicans, more Wave II than Wave I Millennials have contacted a public official (24% vs. 16%).

Wave I Millennial Democrats are also twice as likely as Wave II Millennial Democrats (28% vs. 14%) to have contributed to a campaign; the reverse is true for Republicans where almost two and a half times more of younger Wave II than Wave I Millennial Republicans have contributed to a campaign (29% vs. 12%). Among Independents, the same percentage of Wave I and Wave II Millennials has contributed to a campaign (9% vs. 9%).

The largest gap between Wave I and Wave II Millennials, which ranges from 9 to 29 percentage points, is in the perception of the role of the press as it relates to public affairs. Wave II Millennial Republicans are more than three times more likely than Wave I Republicans to strongly endorse the press's watchdog role, while Wave I Democrats and Independents are more likely than their Wave II counterparts to strongly agree that "it is important for the news media to be a watchdog of powerful people and the government." A similar pattern is found for whether news organizations protect democracy. Younger Wave II Republican Millennials are more likely than older Wave I Republican Millennials to strongly agree news organizations help protect democracy (18% vs. 8%). But among Democrats and Independents, the pattern is reversed. Wave I Millennial Democrats are more likely than Wave II Millennial Democrats (24% vs. 17%) to strongly agree that news organizations protect rather than harm democracy. Similarly, Wave I Millennial Independents (13%) are more likely than Wave II Millennial Independents (4%) to endorse the statement that "news organizations protect democracy more than hurt democracy."

Table 8.6: Wave I and Wave II Millennials in the Voting, Political, Civic, and News Landscape.

	Wave I Millennial Democrats	Wave II Millennial Democrats	Wave I Millennial Republicans	Wave II Millennial Republicans	Wave I Millennial Independents	Wave II Millennial Indepen-dents
Contacted elected official	22%	19%	16%	24%	12%	4%
Donated money to a cam-paign or cause	28%	14%	12%	29%	9%	9%
Frequent attention to public affairs news*	31%	33%	28%	25%	21%	9%
Duty to keep informed*	33%	40%	24%	41%	26%	27%
Feel empow-ered from being informed*	25%	26%	16%	29%	24%	15%
Watchdog press	26%	18%	12%	41%	24%	15%
Help protect democracy	24%	17%	8%	18%	13%	4%

*Differences are not statistically significant.

Race and Ethnicity in the Political Landscape

To the surprise of some, race and ethnicity played an outsized role in the 2016 presidential primary even though it wasn't 2008 when Barack Obama's run for the White House culminated in his election as the first African American president of the United States. Whether because of 2016 Republican presidential primary front-runner Donald Trump's disparaging comments about Mexican immigrants, strident rhetoric about deporting millions of undocumented immigrants, #BlackLivesMatter protests at rallies of Democratic candidates, African American voters in southern states propelling Hillary Clinton to the front of the Democratic primary, or media polls documenting white voters' anger, 2016 will, no doubt, be remembered as a presidential election in which both race and ethnicity took center stage.

While many factors have contributed to the role of race and ethnicity in the 2016 presidential election, the changing demographics of the U.S. population as well as the makeup of the electorate cannot be ignored. Not only is the U.S. population becoming more diverse, voters of color are increasing and white voters are declining. According to the U.S. Census, between the presidential elections of 2008 and 2012, there were an additional 3.6 million voters of color, including 1.7 million black voters, 1.4 million Hispanic voters, and 550,000 Asian American voters. Rather than increase, the non-Hispanic white voter decreased during this same period; between the 2008 and 2012 presidential elections, there were 2 million fewer white voters (The diversifying electorate, 2013).

While the turnout of white and non-white voters and the ultimate impact of race and ethnicity on the 2016 presidential election outcome won't be known until the votes are cast, statistics that capture the relationship between one's political identity from the perspective of race and ethnicity and the connection to news engagement in a mobile landscape can be gleaned from Table 8.7.

In terms of ideology, white Democrats, black Democrats, Hispanic Democrats, and white Independents are more similar than different; the majority are middle of the road or liberal, which is the opposite of white Republicans, who are mostly conservative.

With the exception of Hispanic Democrats, at least three-fifths of each group gets news seven days a week. White Democrats (55%) are most likely to seek news daily and Hispanic Democrats (26%) are least likely to seek news daily, but when it comes to paying attention to public affairs news, there is no significant difference among the five groups. Public affairs news gets the attention of only a minority of each group.

All groups to some degree strongly agree they have a duty to keep informed, with black Democrats most likely and white Independents least likely to strongly agree they have a duty to keep informed. The duty to keep informed no doubt extends to social media, where 14% to almost two-fifths

Table 8.7: Political Party, Race, News, and Mobile

	White Democrat	Black Democrat	Hispanic Democrat	White Republican	White Independent
Ideology and Voting:					
Liberal	43%	35%	32%	2%	17%
Middle of the Road	40%	46%	41%	29%	62%
Conservative	17%	19%	27%	69%	22%
News Engagement:					
Get news 7 days per week	64%	60%	44%	66%	61%
Seek news 7 days per week	55%	40%	26%	48%	47%
High public affairs news attention*	36%	33%	32%	43%	33%
Use mobile device to engage with news	44%	43%	56%	40%	33%
Duty to keep informed	39%	48%	40%	36%	34%
Social Media News Engagement:					
Spend at least 1/3 Facebook time on news	27%	28%	37%	14%	15%
Spend at least 1/3 Twitter time on news	33%	28%	39%	26%	19%
Follow journalists or news organization on Twitter	48%	51%	52%	34%	29%

*Differences are not statistically significant.

Note: Combining race and ethnicity with political party produced 15 different political groups defined by race or ethnicity, but because of small percentages, the following groups were excluded from the analysis reported in Table 8.7: Asian American Democrat, Asian American Independent, Asian American Republican, black Independent, black Republican, Hispanic Independent, Hispanic Republican, Native American Democrat, Native American Independent, Native American Republican.

spend at least 1/3 of time on Facebook and Twitter getting news. Plus, white Democrats, black Democrats, and Hispanic Democrats are more likely than white Republicans and white Independents to follow journalists and news organizations on Twitter. And Hispanic Democrats (56%) are most likely to use their mobile device to engage with news while white Independents (33%) are least likely to use their smartphone or tablet for news.

Trust of Sources of News in the Political Landscape

In the most recent edition of *The Elements of Journalism: What Newspeople Should Know and the Public Should Expect*, authors Bill Kovach and Tom Rosenstiel said, "The quality of our democratic life depends, in short, on the public having the facts and being able to make sense of them" (Kovach & Rosenstiel, 2014, p. 9). But what if the public does not trust the sources providing the facts? That is a valid concern and even more so in the political arena when voters do not trust the sources of the information that will help them make sense of the candidates, their leadership abilities, their stands on issues, and personal attributes before casting a vote. When trust of sources is analyzed by political party, it becomes obvious that the trustworthiness of sources is a factor in voters' ability to get the facts necessary to have a quality democratic life.

Overall, social media are trusted the least, with Twitter, a major platform for distributing news, falling short on the trust scale for Democrats, Republicans, and Independents. According to Table 8.8, only 12% of Democrats, 8% of Republicans, and 6% of Independents trust Twitter a lot. Facebook is trusted a little more than Twitter as a source for news but not by much. Overall, local TV News is the most trusted source for news for Democrats, Republicans, and Independents, with almost one-half (48%) of Democrats trusting local TV news a lot. Compared to Twitter and Facebook, journalists do better on trust but it must be underscored that the trust level is still low, with Republicans the least trusting and Democrats the most trusting. Twenty-three percent of Democrats, 15% of Independents, and 10% of Republicans trust journalists a lot.

Merriam-Webster online dictionary defines politics as the "activities that relate to influencing the actions and policies of a government or getting and keeping power in a government" (Politics, n.d.). Political news, then, would be news about those activities, actions, and polices. While elections, especially presidential elections, are the most covered political activity, politics includes more than elections. In fact, some issues that appear to have little to do with politics may be the most political. Examples include climate change,

Table 8.8: Sources Democrats, Republicans, and Independents Trust for News. % Saying Trust A lot

	Democrats	Republicans	Independents
Local TV news	48%	39%	40%
Local online and print newspaper*	38%	32%	32%
Network TV news	35%	27%	27%
Cable news	33%	23%	23%
National newspaper	31%	23%	22%
Journalists	23%	10%	15%
Government websites	17%	10%	10%
Facebook	16%	13%	12%
Twitter	12%	8%	6%
Bloggers	11%	8%	6%

*Differences are not statistically significant.

what should be in the K–12 curriculum, whether children should be immunized, and even at what age breast cancer screenings should begin.

Because every four years in the U.S., presidential elections are the biggest and longest-running spectacles, featuring outsized personalities and political intrigue, they get more media coverage over a longer period of time than any other event. And now with mobile devices, news coverage has accelerated with voters expecting more frequently updated news so they can get the latest on the candidates and issues. And as the news cycle for updating news in the mobile landscape becomes shorter and shorter, journalists update news more frequently with a tweet or new post to the news outlet's mobile app. Journalists and news organizations are not just trying to satisfy the public's appetite for news, they are also trying to stay ahead of the fierce competition that has also taken up residence in the mobile news landscape.

Every news app includes political news, and that same political news will likely be posted in multiple places on the app: Top News, National News, Politics, and Government. Making the same political news available in multiple sections of the app likely increases the chances that a consumer will find and read the story, but there's no guarantee the story will be trusted. Trust or lack of trust of journalists, political news, or specific news outlets can be barriers to news engagement even when mobile devices make it easy to access news.

Trust of the news media appears to have worsened during the 2016 presidential election, with candidates calling out the media for everything from unfair debate questions to biased coverage and even a failure to cover candidates equally. And as always there's the complaint of too much emphasis on polls and letting polls dictate coverage, or in the case of the 2016 presidential race, letting candidates' issue proposals take a backseat to outrageous statements and controversy. Because in the mobile world we now live in, voters can easily access presidential election news around the clock using their mobile devices, they are more aware than ever as to whether the news media are carrying out their responsibility of reporting on the presidential candidates in a fair and balanced way. If voters feel the news media are failing in that responsibility, trust in the news media, which is already at an all-time low, will not only further erode but the news media will have failed in one of its most important responsibilities: providing voters with the information they need to make an informed decision about whom to vote for.

· 9 ·

DOING JOURNALISM IN A MOBILE WORLD

Since the introduction of the iPhone in 2007, consumers have enthusiastically adopted smartphones, upending entrenched behaviors and attitudes and transforming society into one that is passionate about mobile. This mobile consumer tsunami unsettled journalism, forcing news organizations to rethink and overhaul newsroom culture and routines, create mobile-friendly platforms to deliver news, and invent a mobile-centered business model for making money. Plus, for the first time, news organizations had to put the needs, wants, and interests of consumers first, something they had not done when the old business model was enormously profitable.

News Organizations Adopting Mobile

Although there are differences between adopting an innovation as an individual versus an organization, the mobile-first, mobile specialty, mobile laggard typology introduced in Chapter 1 can provide insight into the mobile adoption process for news organizations. While individual journalists, like consumers, adopted smartphones and tablets for their personal use and even journalistic work, adopting an innovation at the organizational level is a different process that requires a "champion" to be successful. In fact, a champion is so important to the adoption process that *Diffusion of Innovations* guru Everett Rogers (2003) re-affirmed: "The new idea either finds a champion or dies" (p. 414). Rogers defined a champion as a "charismatic individual who throws his or her weight behind an innovation, thus overcoming indifference or resistance that the new idea may provoke in an organization" (p. 414). Champions are extremely important because "anti-innovation champions" lurk in organizations. These opponents of adopting an innovation can "prevent a new technology from reaching the routinization stage of the innovation process" (p. 415).

It helps when the "champion" is the executive editor or news director, according to research, but the champion does not have to hold a high-level position of authority. In the case of mobile device adoption in a newsroom, it may be more important that a champion be mobile-first or a mobile specialist than a top editor. Because the goal of the anti-innovation champion is to stand in the way of change in newsroom culture and routines, that person may be a mobile laggard or non-adopter of mobile devices.

Initiation and Implementation in an Organization's Innovation Adoption Process

Consumer adoption of mobile devices has been so rapid that the normal adoption stages that a news organization might otherwise go through have been truncated. There simply has not been enough time for research and planning prior to implementation. While there are benefits of thorough research and planning, there is also the disadvantage of falling so far behind that it would be virtually impossible to catch up with consumers as they develop routines and preferences with their mobile devices. Consumers have so many things they can do with their mobile devices that accessing news, even if the platform is less than perfect, has to be part of the mobile mix to attempt to have a level mobile playing field.

Mobile Transformation of Newsroom Culture and Routines

Even though iPhones were first sold in 2007, mobile phones were not immediately adopted by newsrooms. The soon-to-take-place mobile transformation of newsrooms wasn't even hinted at in the 2011 documentary *Page One: Inside The New York Times* or HBO's 2012 series, *The Newsroom*. Whether the drama on the screen was real or make-believe, reporters and editors in the *New York Times* documentary and HBO series appeared to be primarily mobile laggards, using their mobile devices for communication.

But today's newsroom culture has undergone a mobile transformation according to the first national survey to examine journalists' smartphone adoption (Molyneux, 2014). Not only had three-quarters of journalists adopted a smartphone but the smartphone adoption appeared to follow *Diffusion of Innovations* theory that explains how adoption of an innovation unfolds in society. When journalists were asked to compare their smartphone adoption with their colleagues, one-third said they acquired a smartphone before their colleagues, 18% acquired a smart mobile device about the same time as their colleagues, 18% adopted a smartphone after their colleagues, and one-fourth had yet to adopt a smartphone (Molyneux, 2014).

Survey results confirmed adopting smartphones was transforming how journalists reported news. Two-fifths used smartphones in their work seven days a week, and smartphone owners were more likely than non-smartphone owners to frequently produce audio and video. Since smartphones have convenient and easy-to-use video and audio recording and editing capabilities, it is not surprising that journalists who have smartphones are using those functions in their work. And because journalists have been required to produce more multimedia since the transformation of the media landscape to digital, it is also not surprising that three-quarters of journalists surveyed used their smartphones to include multimedia in their reporting. What is surprising is that when journalists were asked about the main use of their smartphones for work, almost half (47%) were more like mobile laggards, using their smartphone for communication, including email and texting. One-third of the journalists would be classified as mobile specialists, with 21% using their smartphone for social media and 13% for searching the Internet. Other uses of the smartphone were for photos, video, audio, GPS, and other.

Adopting a smartphone at the individual journalist level is not the same as a newsroom adopting a mobile approach, which can affect newsroom routines from reporting stories to deadlines. With almost half of journalists surveyed saying that their employers required or strongly recommended a smartphone, it is clear that adoption is not limited to individuals; news organizations are also adopting mobile.

Whether a newsroom, overall news organization, or a different type of organization, uncertainty can accompany the adoption process and uncertainty can lead to resistance or adoption failure. According to *Diffusion of Innovations* scholar Everett Rogers: "Some innovations create a high degree of uncertainty in an organization, an uncomfortable state that may foster resistance to the technology" (Rogers, 2003, p. 426). While this may have been the case when newsrooms transitioned from print to digital, this uncomfortable state may have been less of a factor in the adoption of mobile by newsrooms because journalists, just like the general public, were adopting mobile devices for their personal use. But while the simplicity of using an iPhone or other smartphone may have reduced uncertainty, mobile devices have boosted expectations that journalists should always carry their smartphones and be ready to take photos, shoot and edit video, record and edit audio, and tweet and post on social media 24/7. This expectation of the 24/7 mobile multi-media, social media journalist may also be contributing to uncertainty and resistance related to adopting new roles and responsibilities. Still, this expectation of adopting the role of a multimedia journalist may help explain why seven out of ten journalists in the "The American Journalist in the Digital Age" expressed a need for additional training in the following areas that can easily be handled by smartphones: video shooting and editing (30.5%), multimedia (25.6%), photojournalism (18.7%), podcast production (12.4%), mobile and backpack reporting (10.1%), and audio recording (9.3%) (Willnat & Weaver, 2014, p. 23).

Typically, consumers rapidly adopt an innovation because of five attributes: relative advantage, compatibility, trialability, observability, and less complexity (Rogers, 2003, p. 16). The iPhone was perceived to both have an advantage and be compatible with cellphones and BlackBerries, which were the most popular mobile phone when iPhones were introduced. Plus, because of the Apple retail stores, consumers could try the phones, which scored high on the adoption attribute. And because of the coolness factor, iPhones were shown off to friends and family. The elegant simplicity of the iPhone easily made the case for "less complexity."

While these five attributes were relevant to the rapid adoption of smartphones by consumers, there were benefits to using smartphones that no doubt played a role in newsroom managers requiring and encouraging journalists to use smartphones in their work. One-third of journalists (33%) participating in the national survey said smartphones enabled them to work from anywhere. Twenty-eight percent said smartphones enabled them to keep in touch while 21% said smartphones enabled them to make immediate updates. Fourteen percent emphasized the convenience of smartphones and 5% said smartphones made multimedia possible.

Journalism Principles in the Mobile Landscape

Smartphones have given journalists more flexibility and opportunities to keep in touch and update news but they have not changed the fundamental principles of journalism. In fact, the "window on the world" metaphor put forward by sociologist Gaye Tuchman, author of *Making News: A Study in the Construction of Reality*, is just as relevant in today's mobile news landscape as it was in 1978 when the only platforms for news were TV, print newspapers, radio, and news magazines. Tuchman's definition of news was consumer-centered, unlike Carl Bernstein of Watergate fame, who defined news while speaking to journalism and communication students and faculty at the University of Texas at Austin, April 21, 2011. Tuchman said, "News aims to tell us what we want to know, need to know, and should know" (1978, p. 1), while Bernstein defined news as "what journalists say it is." While in 1978, Carl Bernstein's definition was more likely to be true, in today's social media–mobile world, with citizen journalists, "accidental" journalists, and user-generated content posted on social media platforms, telling us what we want to know, need to know, and should know is no longer the exclusive domain of journalists.

Much of the traditional criteria that journalists have used to judge news value—audience, impact, proximity, timeliness, prominence, unusualness, conflict, visual and audio impact—are also used for judging news value in the mobile news landscape. But because of small mobile screens some criteria such as prominence, conflict, and visual impact can lead to distortion in proportion and bias that excludes and stereotypes specific groups.

The audience news value can also lead to distortion when the goal of principled journalism is twisted to focus on attracting the maximum audience. In fact, in 1947, the Hutchins Commission on the Freedom of the Press criticized the press for placing more value on attracting the maximum audience than meeting the requirements a free society should expect of its press. Almost seven decades later in a mobile-social media world, a similar criticism can be made of the press when analytics are substituted for news judgment. Just because the analytics capability is available, it does not mean decisions about what news to cover, what headlines to write, what videos to produce, or what developing news to tweet should be based on generating the most clicks. An analytics-first approach does not serve journalism, consumers of news, or the public good.

Concerns about journalism, its role in society, and how well journalism serves the public good are not unique to the mobile-social media world that journalism operates in today. This concern has been present throughout journalism's history. In addition to criticisms from the 1947 Hutchins Commission Report, two decades later in 1968, the Kerner Commission criticized the news media for its diversity failures in both reporting news and newsroom employment. Another three decades passed before 25 journalists began looking inward to figure out why journalism was damaging the public interest rather than serving it (Kovach & Rosenstiel, 2014). The results of those initial discussions, as well as surveys, depth interviews, focus groups, and forums around the country, led to the publication of *The Elements of Journalism*, which is now in its third edition. While all journalism elements are relevant, the five italicized elements require special attention if journalism is to serve the public interest in a news landscape that has been transformed by mobile devices and social media.

1. *Journalism's first obligation is to the truth.*
2. *Its first loyalty is to citizens.*
3. *Its essence is a discipline of verification.*
4. Its practitioners must maintain an independence from those they cover.
5. It must serve as an independent monitor of power.
6. It must provide a forum for public criticism and compromise.
7. *It must strive to make the significant interesting and relevant.*
8. *It must keep the news comprehensive and in proportion.*
9. Its practitioners have an obligation to exercise their personal conscience.
10. Citizens, too, have rights and responsibilities when it comes to the news.

(First nine elements published in *The Elements of Journalism: What Newspeople Should Know and the Public Should Expect*, by Bill Kovach & Tom Rosenstiel, 2001; 10th element added to updated and revised edition published 2007.)

The reports and book produced from the three national critiques of journalism may be the most well-known but they are not the only criticisms of journalism. In fact, criticisms of journalism are year-round and ongoing by the minority press, journalism associations, news organization ombudsmen, classes taught in journalism schools, textbooks, conference papers, research published in academic journals, media and public opinion surveys, and even presidential candidates. Unfortunately, these press shortcomings, especially as they relate to coverage of racial and ethnic groups, women, and Millennials, can be made worse in the mobile landscape when analytics substitute for news judgment and formatting of stories and photographs on small smartphone screens distort the news by excluding, stereotyping, and failing to keep news in proportion.

Journalism Ethics in the Mobile Landscape

Long before there was a mobile landscape, journalism developed ethics codes to make the field more professional. In fact, the first ethics code dates back to 1922 when the American Society of Newspaper Editors (ASNE) adopted what they initially called "Canons of Journalism" and a half century later renamed, "Statement of Principles" (n.d.). The ethics code emphasized responsibility, freedom of the press, independence, sincerity, truthfulness, accuracy, impartiality, fair play, and decency. Since the first journalism code of ethics, virtually every journalism association, whether for editors, photojournalists, broadcast journalists, managing editors, or journalists in general, has developed and adopted an ethics code emphasizing the ethical standards codified in 1922. While ethical standards are constant, the codes themselves are generally revised to incorporate changing times, a changing media landscape, or new ethical dilemmas journalists encounter. A good example of change is the Code of Ethics for the Society of Professional Journalists (SPJ). Revised as recently as September 6, 2014, SPJ's ethics code now contains four guiding ethical principles (SPJ, 2014): 1. Seek truth and report it; 2. Minimize harm; 3. Act independently; 4. Be accountable and transparent. Each ethical principle is followed by ethical behavior that should be exhibited during the process of journalism. Some of this behavior speaks to the elements of journalism and some applies to conducting journalism in the mobile and social media landscape.

For example, to achieve the first principle of seeking and reporting truth, the code lays out 18 actions that journalist should do; 6 of those actions speak to criticisms of the press:

- "Provide context."
- "Gather, update and correct information throughout the life of a news story."
- "Give voice to the voiceless."
- "Boldly tell the story of the diversity and magnitude of the human experience. Seek sources whose voices we seldom hear."

- "Avoid stereotyping. Journalists should examine the ways their values and experiences may shape their reporting."
- "Never deliberately distort facts or context, including visual information."

To achieve the second principle of minimizing harm, the code has seven actions for journalists, and two speak directly to criticisms of journalism:

- "Balance the public's need for information against potential harm or discomfort."
- "Show compassion for those who may be affected by news coverage. Use heightened sensitivity when dealing with juveniles, victims of sex crimes, and sources or subjects who are inexperienced or unable to give consent. Consider cultural differences in approach and treatment."

The third ethical principle, "act independently," specifies five actions that address avoiding conflicts of interest, paying for news, and favored treatment. One action demands vigilance because it can be less obvious in a mobile format: "Distinguish news from advertising and shun hybrids that blur the lines between the two. Prominently label sponsored content."

The fourth and final principle of SPJ's revised Code of Ethics, "Be accountable and transparent," contains five actions that are critical if the public is to view the press as ethical and trustworthy:

- "Explain ethical choices and processes to audiences. Encourage a civil dialogue with the public about journalistic practices, coverage and news content."
- "Respond quickly to questions about accuracy, clarity and fairness."
- "Acknowledge mistakes and correct them promptly and prominently. Explain corrections and clarifications carefully and clearly."
- "Expose unethical conduct in journalism, including within their organizations."
- "Abide by the same high standards they expect of others" (SPJ, 2014).

Ironically, the press has excelled at the action of exposing unethical conduct in journalism, including within their own organizations. The list is long and because of the prominent display by individual organizations and their competitors, the public may have a perception of excessive unethical behavior, which in reality is rare and not the norm.

Even so, unethical behavior by journalists has ensnared some of the most prestigious news organizations dating back at least to the 1980s. In 1981, Janet

Cooke of the *Washington Post* wrote a story about an eight-year-old heroin addict named Jimmy, which won a Pulitzer Prize. But the Pulitzer had to be returned because Janet Cooke had fabricated the story, titled "Jimmy's World" (Cooke, 1980). Four years later, in 1985, *The Wall Street Journal*'s Foster Winans was convicted of fraud and conspiracy for leaking the contents of his articles prior to publication to two New York stockbrokers, who earned almost $700,000 from insider trading on the information (Lewin, 1986).

In November 1992, NBC's *Dateline* broadcast a story allegedly exposing defective gas tanks in GMC trucks. It turned out the exploding gas tank captured in the story was a result of the producers secretly rigging the truck so it would explode during the crash that had been staged (Parrish & Nauss, 1993).

The film *Shattered Glass* tells the real-life fraud of *New Republic* associate editor Stephen Glass, who was fired in 1998 for making up half a dozen articles and inventing portions of more than 20 others that were published in the elite national magazine (Bissinger, 1998; Noer, 2014). Brian Walski, a photojournalist for *The Los Angeles Times*, who was covering the Iraq War combined two separate photographs into one image and was fired for failing to uphold a fundamental photojournalism ethical standard (Irby, 2014).

Perhaps, the most notorious ethical breach in journalism was by 27-year-old *New York Times* reporter Jayson Blair, who plagiarized quotes, fabricated scenes, and faked interviews (Barry et al., 2003). Not only was Jayson Blair fired but so were the executive editor and managing editor (Steinberg, 2003). In 2004, *USA Today* star foreign correspondent Jack Kelley was also fired after "fabricating substantial portions of at least eight major stories" and plagiarizing "nearly two dozen quotes or other material from competing publications" (Steinberg, 2004).

And in 2015, the No. 1 network news anchor Brian Williams was suspended for six months without pay from *NBC Nightly News* for exaggerating his experience in a helicopter attack while covering the Iraq War (Steel & Somaiya, 2015). After his six-month suspension, Williams was re-assigned to a newly created job of anchoring breaking news and special reports at NBC's cable news network MSNBC (Kludt, 2015).

The line between unethical journalism and poor reporting is sometimes blurred, but there are differences, even though from the public's perspective the difference may not matter. But some poor reporting is so egregious that it is as bad as the worst ethical infractions and forever shakes the public's confidence in the press. Whether unethical or poor reporting, both get extensive coverage in the press, such as the *Rolling Stone* magazine gang rape story that failed to follow fundamental reporting techniques, or *The New York Times*' Judith Miller, whose misleading and inaccurate stories about the Iraq War, including reports about weapons of mass destruction, were published on the newspaper's front page (Coronel et al., 2015; Bill Moyer's Journal, 2007). The *Rolling Stone* piece was retracted and Judith Miller's stories, as well as senior editors, were publicly criticized by *The New York Times* Public Editor, making headlines and further eroding the public's trust in journalism (Okrent, 2004).

Still, such major journalism blunders are not the norm, but they look like the norm when news organizations report on them. Whether unethical or poor reporting, both must be promptly and prominently corrected as specified by the SPJ Ethics Code, but they also must be put in context. Plus, journalism and the public good would benefit if the opportunity were turned into a teachable moment where journalism balanced its critique with explanations of the principles and process of journalism, with special emphasis on the lengths journalists will go to verify information in order to report accurately.

That's why *All the President's Men* should be recommended viewing for the public. The 1976 film shows how *Washington Post* reporters Bob Woodward and Carl Bernstein investigated Watergate and meticulously verified information that uncovered criminal acts in the White House, eventually leading to the resignation of President Richard Nixon. *Spotlight*, which is being compared to *All the President's Men*, is also a lesson on how journalists report the news (Hornaday, 2015). The 2016 winner of the Academy Award for Best Picture portrays the *Boston Globe*'s "shoe-leather investigation" of the "decades-long scandal of child sexual abuse within the Catholic Church, involving 70 local priests, more than 1,000 victims, and the collusion of Cardinal Bernard Law in covering up the crimes" (Hornaday, 2015).

While these films are excellent teaching tools, the majority of the public has likely not seen these examples of how journalism should be done. Unfortunately, the public may have seen the disturbing 2014 film *Nightcrawler*, an example of journalism at its worst. Journalism cannot control how journalism is portrayed in films or on TV or even on what is said about it on the presidential campaign trail, but if there is any hope of increasing public trust, journalists and news organizations must be pro-active in helping the public understand what journalism stands for, what journalism excellence looks like, and how it is practiced.

While SPJ's most recent revision of its ethical code contains issues that are relevant to today's mobile news landscape, the Online News Association (ONA), since 2013, has been developing an ethical code explicitly for the digital landscape. This code is unusual because it recognizes that all journalists are not the same and may very well have different goals for their news platforms. This recognition of differences emerged from the reality of the journalism profession that all journalists today did not come out of the "objective" legacy news media model of the past or take the path to the journalism profession through journalism schools.

In recognition of different journalistic approaches in the digital news landscape, the Online News Association created the Build Your Own Ethics Code Project (2015) that encourages ethics while recognizing and respecting differences across the field. Spearheaded by Associated Press Deputy Managing Editor/Standards Editor Tom Kent, the Build Your Own Ethics Code Project addresses approximately two dozen topics, including racial, ethnic, and gender references, user-generated content, use of content from social media accounts, data journalism, and online commenting. Whether ethical codes emerged from legacy media or the different platforms in today's news landscape, there's recognition that even if customized, an ethical code in some form or fashion is essential in the world of mobile journalism.

Social Media in Mobile Journalism

Before smartphones emerged in the media landscape, social media was already transforming journalism beyond news websites. Although the most used social media sites, Facebook and Twitter, were born on desktop computers, their mobile transformation has influenced how journalism is being conducted. In fact, when journalists who participated in "The American Journalist in the Digital Age" survey were asked the impact of social media on their work, it was clear that social media's impact was not limited to reporting (Willnat & Weaver, 2014). Social media have not only enabled journalists to brand and market themselves in ways that were unthinkable in the past but smartphones also have made personal branding and marketing possible around the clock. In fact, according to Table 9.1, fourth-fifths of journalists agreed or strongly agreed that social media have enabled them to promote themselves and their work.

Social media have also enhanced journalists' relationships with the audience, something that is critical to the future of journalism. Seven out of ten journalists agreed or strongly agreed that social media have made it possible to engage with the audience and almost one-third said their credibility has been enhanced as a result of social media (Willnat & Weaver, 2014).

Table 9.1: Impact of Social Media on Journalism from "The American Journalist in the Digital Age" Survey.
% of Journalists Who Agree or Strongly Agree

Brand and Promote:	
Promote journalist and work	80%
Enhance Relationship with Audience:	
Engage with audience	69%
Enhance credibility	30%
Improve Reporting:	
Faster reporting	62%
Communicate better with relevant people	49%
Cover more news	29%
Improve productivity	25%
Decrease workload	6%

Note: Compiled from "The American Journalist in the Digital Age," (Willnat & Weaver, 2014).

Other benefits of social media are specifically related to reporting and are likely enhanced by the use of smartphones. More than three-fifths of journalists agreed or strongly agreed that social media have contributed to faster reporting and 29% agreed or strongly agreed that social media make it possible to cover more news (Willnat & Weaver, 2014.) While one-quarter felt social media improved productivity, it wasn't surprising that only 6% said social media decreased their workload. The impact on better communication with relevant people that social media made possible has no doubt improved even more with the adoption of smartphones as a reporting tool.

While social media have enhanced journalism, smartphones have significantly enhanced social media's impact on journalism. Not only do social media apps make it easy for consumers to access social media, the social media apps also make it quick and easy for journalists to post stories on social media. In fact, journalists can tweet about a story that is in the works and use social media to break news. Plus, apps make it very easy for journalists who use social media as a source for gathering news.

The impact of social media on journalism has been so significant that 28% of journalists responding to "The American Journalist in the Digital Age" survey expressed a desire for more training on social media engagement (Willnat & Weaver, 2014). Additionally, because social media raise new issues for newsgathering, news sources, disseminating news, journalism routines, standards, and ethics, news organizations and associations have incorporated rules for using social media into their style guides. And with the integration of social media into mobile, opportunities to enhance journalism and reach and cultivate mobile-first consumers and mobile specialists increase, but potential problems unique to using social media to do journalism can further erode trust in journalists and news organizations.

In recognition of the fact that today social media are not just a platform to disseminate and access news but also a tool for gathering news, the Associated Press (2013) has added social media guidelines to its *Stylebook* for journalists. First and foremost, the social media guidelines emphasize that journalistic ethics and standards still apply even though the medium is not a traditional news platform for disseminating news or a tool for gathering news. The AP guidelines caution that "social networks should never be used as a reporting shortcut when another method, like picking up a phone or knocking on a door, would yield more reliable or comprehensive information" (The Associated Press, 2013, p. 370.), and when social media are used as a source of news generated by users, journalists have to evaluate the content using the highest journalism standards, beginning with verification.

> User-generated content, or UGC, is the term commonly used in the news industry for content with news or editorial value that has been produced by non-journalists. Sometimes this is also referred to as citizen journalism, when members of the public capture news events on their own devices either by chance or by purposely pursuing a story. User-generated content may be found via social networks but could equally be given to a reporter on the scene of a news event. (The Associated Press, 2013, p. 387)

Social Media Gatekeeping in the News Landscape

Perhaps as important as the opportunity to have access to content that social media users post and share is an expanded gatekeeping role for journalists. It was 1950 when David Manning White (1950) applied "gatekeeping" to editorial decision making about what gets published and what gets rejected. Over the past decades, journalism scholars have applied gatekeeping to what gets broadcast and rejected in TV and online news, and researchers have examined women editors to see if women make different gatekeeping decisions from men about what is published (Berkowitz, 1990; Singer, 2001; Cassidy, 2006; Bleske, 1991). With social media now an integral part of the news landscape, journalists are now social media gatekeepers, deciding what gets posted on Facebook and what gets tweeted and retweeted on Twitter. In the original 1950 gatekeeping study when the editor, who was called "Mr. Gates," provided a rationale for news from the wire service that was selected to be published or rejected, the audience was often cited as a rationale for editorial decisions (White, 1950). While "Mr. Gates" was one wire editor, it must be underscored that there are many gatekeepers throughout the life of a news story or news cycle: "In the production of a news story, there are multiple gatekeepers, including reporters, photographers, videographers, section editors, managing editors, producers, news directors, and their deputies, to name a few" (Poindexter, 2012, p. 121). And for watchdog investigative pieces that pit the "Fourth Estate," with its First Amendment protection, against the Executive Branch of the U.S. government, the executive editor and publisher may be the final gatekeeper arbiters.

In addition to multiple gatekeepers, there are multiple factors that have been identified as influencing gatekeeping decisions: news values and professional standards; formal education and training; newsroom policies, culture, and routines; newsroom staff racial, ethnic, gender, and generational composition; unconscious assumptions about the subject (Shoemaker & Reese, 2014; Willnat & Weaver, 2014; Poindexter et al., 2008; Poindexter, 2012). And in the case of social media gatekeeping, analytics, marketing, and trending potential should be added to the list of influences on gatekeeping.

In social media gatekeeping, it should also be stressed that some of the traditional gatekeeping influences may not apply because of the distinction between the institution and individual. The news organization will have a Facebook page and accounts for Twitter, Instagram, Snapchat, and other social media sites, while an individual journalist will have his or her own Twitter account. In the case of the institutional social media accounts, the list of influences may play a role in what gets posted, but for individual journalists, the journalist is the sole gatekeeper, except in the case when news organizations establish policies for how journalists represent themselves on social media and what they can and cannot do. The Associated Press, for example, has very specific polices about what AP journalists can do on social media. The five pages of social media policies are "designed to advance the AP's brand and staffers' personal brand" but staffers are not allowed to "express personal opinions on controversial issues of the day" (The Associated Press, 2013, p. 394).

With social media gatekeeping, there are, of course, ethical considerations at both the institutional and individual level, which is why ONA's Build Your

Own Ethics Code Project addresses the ethical implications of retweeting, identification of self, sharing news developments, including rumors and hearsay, friending and following, and errors and corrections. As shown in Appendix E, ethical approaches to these issues may be influenced more by whether the news organization strives to be impartial or more opinionated.

Mobile Gatekeeper

Just as there are issues unique to social media gatekeeping, there are issues unique to news on mobile. Some of the issues are specific to gatekeeping on mobile; other issues are related to framing, a theory first applied to journalism in the early 1990s (Reese et al., 2001). Framing as a theory applied to journalism has not been around as long as gatekeeping, but the concept has always been intertwined with gatekeeping because a journalist is not just deciding what to publish or exclude, the journalist as gatekeeper is also determining the frame or *how* the story will be reported:

> A frame is a central organizing idea for news content that supplies a context and suggests what the issue is through the use of selection, emphasis, exclusion, and elaboration. Much of the power of framing comes from its ability to define the terms of a debate without the audience realizing it is taking place. Media framing can be likened to the magician's sleight of hand—attention is directed to one point so that people do not notice the manipulation that is going on at another point. (Tankard et al., 1991)

And because of the small screen on a smartphone, there's a greater chance that the frame through which the story is told excludes or over-emphasizes in a way that can undermine journalism principles and distort what otherwise might be a "representative picture" that the Hutchins Commission said society had a right to demand of its press.

Just as news organizations have guiding principles, routines, and staff for presenting the news on a website or during a broadcast, there are guiding principles and routines for posting news on a mobile news site or app. The layout on a smartphone is generally vertical where a consumer can tap, scroll, and swipe through the various sections, including top news, national, sports, business, entertainment, technology, health and science, or art. Within each news section, there are screens framing stories with a headline and thumbnail photo, and by tapping on a story of interest, the first screen of the full story with a headline, larger photo, and lead will emerge.

Figure 9.1 reminds us that while journalists may concern themselves with their individual news stories, news organizations have to consider the full mobile device because each and every app that resides on a mobile screen is a competitor for consumers' time and attention. For example, the mobile screen in Figure 9.1, which is one of six screens from my iPhone, displays 13 news apps and 2 social media apps. Another nine news apps and four additional social media apps, including Instagram, Snapchat, LinkedIn, and Pinterest, can be found on other screens. The YouTube video-sharing app as well as Periscope, which makes live streaming as easy as shooting video, also have a home on one of my mobile screens. From "Watch ESPN," I can live stream sports, which is a necessity for me if I'm not home during Wimbledon, the U.S. Open, Australian Open, or French Open. And Hulu and NetFlix apps make it possible for me to watch movies and original and network TV shows while I'm riding the stationary bike in the gym. Mixed in between news and social media apps, I have Google and apps for music, banking, games, podcasting, exercise, weight lifting, and even meditation that I downloaded after reading an article on one of my news apps.

Figure 9.1: Smartphone Screenshot of News and Social Media Apps.

While Chapter 5 introduced news innovations such as BuzzFeed and NYT Now that appeal to Millennials, there have been news innovations by social media and tech companies that should resonate with mobile consumers who have at least some interest in news: Snapchat's Discover, the Apple News app, Twitter Moments, and Facebook's Trending. Make no mistake; these news innovations produced by non-news companies also compete with journalists and news organizations that are doing journalism in a mobile world.

News apps make it convenient to access news but they are not the only way to access news on a mobile device. Consumers may get news through social media apps or browsers. When notifications are turned on, news can break through the locked screen. As recalled from Chapter 4, breaking news is one of the most popular types of news and mobile devices, whether through notifications or banners on apps, make it easily available to consumers.

Because of the rapid adoption of mobile devices, news organizations did not have any choice but to make news available on mobile so they would not be left behind. This forced adoption has not always led to providing mobile news in a format that is best from the consumer perspective. In some cases, news organizations are simply automatically shoveling the news from the website to the mobile site or app. In other cases, news organizations are trying to develop a mobile news product with the consumer in mind. Consumers, especially mobile-first and mobile specialists, are passionate about their mobile devices but why are so few consumers passionate about news on mobile? Is it because news organizations haven't thoroughly rethought and overhauled newsroom culture and routines to make the news they produce mobile-first? Or could it be news organizations are unaccustomed to putting the needs, wants, and interests of consumers first? Or maybe news organizations have not fully embraced mobile-first because the path to profitability is long and winding and impediments along the way appear overwhelming and insurmountable. These issues and more will be addressed in Chapter 10.

· 10 ·

THE FUTURE OF NEWS IN A CROWDED MOBILE LANDSCAPE

When the first Millennials were born in the early 1980s, legacy news media dominated. The *New York Times*, the *Washington Post*, and the *Los Angeles Times* were not only premier awarding-winning sources of news, their storied histories and influence were chronicled in best-selling books and studied in journalism classrooms across the country. CNN was in its infancy, *USA Today* had just launched as a general interest national newspaper that carried color on its front page, and the only people aware of a computer network that would one day be called the Internet were computer scientists in academia and Department of Defense workers.

Even though there were 1,701 U.S. newspapers, with a daily circulation of almost 63 million, when the first Millennials were born the daily reading of newspapers was no longer an institutionalized behavior and most adults turned to TV as their primary news source (Newspaper circulation, n.d.). Just as no one could have predicted the rapid transformative impact of the iPhone when it was unveiled in 2007, few, except Steve Jobs, could have predicted the sweeping changes the Macintosh would bring about when it was unveiled in 1984, just one year after the first Millennials were born. "Jobs knew when he unveiled the Mac that it would propel the personal computer revolution by being a machine that was friendly enough to take home" (Isaacson, 2014, p. 365). Because the Mac was "friendly enough to take home," it was rapidly adopted by homes and schools and became a part of the everyday lives of Millennials as they grew up.

By the time the first Millennials became pre-teens, the Internet had been "made accessible to everyone" (Isaacson, p. 401) and America Online (AOL) had become the gateway through which families accessed the Internet. "With America Online, you didn't have to be an academic or a Department of Defense techie to get on the Internet. You could be a kid" (Poindexter, 2012, p. 2).

The connectedness that America Online made possible for families in the 1990s, was re-imagined for Millennial college students when Facebook was launched in 2004 from a desktop computer in a Harvard dorm room. Three years later when the iPhone was unveiled, Facebook, which at first was restricted to college students, had been opened to everyone 13 years and older. During the 1990s, after having conducted a variety of experiments, news organizations began making news available through consumer-oriented online services such as America Online (Weiss, 2008). Newspapers also built their own brand on the Internet by constructing news sites made possible because of the World Wide Web, created by Tim Berners-Lee who "insisted that the Web protocols should be made available freely, shared openly, and put forever in the public domain" (Isaacson, 2014, p. 413). The addition of consumer-friendly browsers made it possible to navigate the Web and the news sites that were populating it.

Much had changed in the news landscape from the time the first Millennials were born to the time that the first Millennials graduated college. Alex typifies the evolution of the Millennial news consumer growing up in a news landscape undergoing change. Alex grew up in a home in which news was a priority. Her parents subscribed to two newspapers—one local and one national—and they watched the local and network news every day, morning and evening. Alex's mother also listened to NPR's *Morning Edition* and *All Things Considered* when she drove Alex to and from school. While Alex's mother was first introduced to a computer at work, Alex was first introduced to a computer at home. When her mother bought a desktop computer for her home office, she loaded it with family-friendly America Online for access to the Internet, so from the time Alex was in elementary school, she could dial up the Internet through America Online. With America Online, you didn't have to search for news because news welcomed you and so did email. Alex used America Online to do research for school and Instant Messaging replaced the phone in how she talked to her friends.

Although her parents were news enthusiasts, Alex wasn't a big fan of news growing up. She thought news was boring, but that changed when she went off to college at a university in the Midwest. Alex credits *USA Today*'s Collegiate Readership Program with her newfound interest in news. Local, regional, and national newspapers were free to everyone with a student ID. Alex's parents bought her a cellphone when she was just 14, so from her early teenage years, a mobile phone was always an integral part of her life. That first cellphone was for communication, so Alex could call her parents to pick her up following afterschool activities. Unbeknown to her parents until the bill came in, Alex discovered texting, which she used to communicate with her friends.

Throughout college, Alex's fashion sense extended to her choice of cellphone and when she graduated college, she also graduated to a Blackberry, which at the time was the most popular mobile phone on the planet, even though it was fundamentally a device for communication. Alex's mother was the first in the family to buy an iPhone—not because of the features but because of the large type. So Alex's adoption of an iPhone was like an adoption theory case study. The iPhone scored high on each characteristic necessary for adoption: 1. Relative advantage; 2. Compatibility; 3. Complexity (that is, simplicity); 4. Trialability; 5. Observability.

The iPhone didn't just have a relative advantage over the Blackberry; it was far superior because there was nothing it couldn't do and do it simply and easily and elegantly. It even played music, which meant Alex would no longer need to carry her iPod. Plus, when Alex was home from graduate school for Christmas vacation, every time her mother put down her iPhone, Alex picked it up and played with it, so she wasn't just observing, she was trying it out and loving it. So Alex's mother bought her an iPhone and Alex used it for everything, from social media to listening to music to researching information to engaging with news and texting. There was little that Alex couldn't and didn't do with the iPhone. She was one of the first mobile-first consumers.

The speed of the adoption of iPhones and other smartphones was so rapid that it seemed more like a spontaneous consumer-led revolution than an evolution, planned and implemented by corporate and newsroom executives. The old way of going online through desktop and laptop computers was replaced by a simple, elegant device that fit in a pocket or purse. The device was cool and it was fun and consumers adored it. The device was even utilitarian. You didn't have to search for a flashlight, a stopwatch, timer, calculator, or map because it was all in one sleek device. Communicating, whether by phone call, texting, email, or video chat, was always conveniently available. You didn't even have to remember a phone number or email address because contact information was stored on the device and retrieved with a finger tap. Need to set your alarm clock or check the weather? Want to take a photo or video? All available in one simple device with a finger tap or two. And in one seamless maneuver, you could edit the photo or video and store it in an album or share it with friends and family.

Smartphones also made it easy and convenient to get news, which may be why Alex now gets news every day. Alex doesn't download apps on her iPhone because she says they take up too much space but she has "liked" the Facebook page of the *New York Times*, NPR, *USA Today*, CNN, and local news outlets so she sees their news on her Facebook News Feed. On her iPad Mini, Alex has downloaded the free *New York Times* app, NYT Now, that was perfect for Millennials. When the app charged a subscription, Alex never would have downloaded it even though she was aware of it. After Apple added Apple News as part of its software update, Alex discovered it and added it to her sources of news because of the convenience—the app was visible on her smartphone screen—and its presentation of news was as enticing as NYT Now. Alex's news diet is not limited to her smartphone—she also listens to NPR's *Morning Edition*, *All Things Considered* and *Here and Now* while driving to and from work.

As a Millennial, Alex isn't typical as far as her high level of interest in news but she is a reminder that through socialization and education and individual initiative and convenience and easy access to news on smartphones, news consumers can be developed. But news organizations need to do more and they need to do it in a hurry. It is simply not enough to develop apps, make news available on social media or Apple's news app, or decide what should be posted as news by its click potential.

News organizations took more than a decade to become online-first but because of the rapid adoption of smartphones, news organizations have had little time to figure out mobile. And what has made mobile more challenging for news organizations is the direct competition on the mobile screen for consumers' time and attention. Why Facebook, Netflix, or "Words with Friends" and not news?

And what about the business model for mobile? The newspaper business model had already imploded online. Would consumers pay for news on mobile? What about advertisers? So many advertisers had already been lost to the Internet and would there be a chance of their returning if news is not a priority in the lives of the largest living generation?

News organizations, especially newspapers, had already proven they were slow to react to the Internet challenge. And they haven't been the swiftest in reacting to the mobile challenge. If news organizations are going to compete and remain relevant in the mobile landscape, they have to think mobile-first and be willing to once again disrupt the way things have been done in the past.

But changing newsroom routines to be mobile-first is not the only challenge news organizations face in the mobile news landscape. Newsrooms also have to reverse declining news consumption; adopt a mobile approach that produces great journalism; strengthen the public's belief that being informed is important; build the public's trust in the press; and develop a viable (robust) mobile news business model.

Reversing Declining News Consumption

There's no definitive evidence that the erosion of getting and seeking news across generations is due to the transformed news landscape, but there are enough clues to support my theory that the news consumption decline displayed in Table 10.1 is due to a combination of factors, including a tradition

Table 10.1: Comparing Getting and Seeking News 7 Days a Week by Baby Boomers, Generation X, and Wave I and Wave II Millennials.

	Baby Boomers Born 1946–1964	Generation X Born 1965–1982	Wave I Millennials Born 1983–1991	Wave II Millennials Born 1992–1999	Post-Millennial Generation Born 2000 and Beyond
Get news 7 days a week	75%	50%	40%	33%	?
Seek news 7 days a week	60%	37%	28%	17%	?

of legacy news organizations putting consumers *last*, a failure to recognize that Millennials are worthy of being included in the news, the emergence of social media created by Millennials for Millennials that court them on cool, always-present mobile devices, a decline in socialization of upcoming generations to believe that being informed is important, and a lack of confidence in the news media that accuracy, truth, balance, inclusiveness, and unbiased reporting are not just essential but paramount.

Whether it's getting news or seeking news seven days a week, the percentage difference between Baby Boomers and the youngest Millennials is about the same according to the News for a Mobile-First Consumer Survey: 42% and 43%. As the youngest Millennials finish high school and go to college or to work, what will be the relationship between getting and seeking news and the post-Millennial generation that follows them? A further decline or an increase? While this question is being asked about news engagement, it is not being asked about social media and mobile devices. There's no doubt that the post-Millennial generation will embrace social media and smart mobile devices. And they will adopt wearables and smart products that have yet to be imagined.

So what, if anything, can news organizations do?

Millennials, News, and Social Media: Is News Engagement a Thing of the Past? (Poindexter, 2012) documents Millennial criticisms about the news media and proposes more than two dozen "Best Practices for Covering and Engaging the Millennial Generation" displayed in Figure 10.1. Unfortunately, four years after these recommendations were proposed, many are still relevant today. What is particularly unfortunate is the first "Best Practice" of regularly including Millennial stories in news coverage is still not taken seriously. There has been some improvement but much more still needs to be done if there is any hope of putting the brakes on the news engagement decline.

Some news organizations have interpreted including Millennials in news coverage as publishing separate Millennial publications or establishing separate Millennial sections on their news sites. Those efforts have had limited success in the past and will likely have little success in the future because they don't represent inclusion; they artificially segregate news by so-called Millennial topics. Millennial perspectives, sources, and even topics that might be of interest to Millennials should be integrated throughout the whole site. Further, news that Millennials should know should be written so it's clear why the story matters to their lives.

Figure 10.1: Best Practices for Covering and Engaging the Millennial Generation.

- ✓ Regularly include Millennial stories in news coverage.
- ✓ Include Millennials as sources in general news and trend stories that impact Millennials and are of interest to them.
- ✓ Stories about Millennials should include explicit references to this generation in the headline.
- ✓ Explicit Millennial stories should include Millennial sources and these Millennial sources should be among the first quoted. Whether Millennial explicit or implicit, care must be taken to include direct Millennial quotes and not just paraphrased Millennial voices.
- ✓ Reduce stories about Millennial celebrities (entertainers, athletes, actors, reality show participants); increase stories about non-celebrity Millennials contributing to society and making a difference.
- ✓ Balance crime coverage with stories that are not about Millennial criminals.
- ✓ Include Millennial voters and issues in election stories.
- ✓ Not all Millennials are the same so strive for diversity in stories and voices.
- ✓ For stories that are important for Millennials to know, include the reason the story matters, perhaps in a sidebar, but don't insult Millennials by "dumbing down" the story.
- ✓ Use neutral tones in news coverage; avoid snide, disrespectful tones.
- ✓ Make headlines and photo captions explicit in stories about Millennials.
- ✓ Use multimedia strategically and effectively to showcase Millennial stories.
- ✓ Use linking effectively to provide context and facilitate engagement.
- ✓ Tag Millennial stories with key words: Millennial, young adults, 20-somethings, youth, Generation Y, GenY.
- ✓ Promote and archive stories about Millennials, stories of interest to Millennials, and stories Millennials should know.
- ✓ Generational rites of passage are a great source of Millennial story ideas. These rites-of-passage stories should be handled with sensitivity and from the Millennial point of view rather than journalists', parents', or employers' perspectives. Tag, promote, and tweet these stories so that Millennials can find them when they search for information to help them navigate these critical passages.
- ✓ Make improved Millennial news coverage a priority by making it the responsibility of an editor who reports directly to the newsroom's chief news executive.
- ✓ Regularly audit news content to assess Millennial inclusion; make sure that stories of interest to Millennials and stories that Millennials should know are written so Millennials can better relate to them.
- ✓ Integrate assessment of Millennial news coverage, with special attention paid to inclusion, story type, sources, and tone, into newsroom culture and routines.
- ✓ Make it a priority to assess how Millennials of color, women, and young voters are handled in the news. On an ongoing basis, assess if they are included or ignored, stereotyped, or treated as second-tier. If coverage is consistent with concerns identified in Chapter 5, develop and implement a plan that will improve coverage of these three key groups.
- ✓ Appoint a Millennial advisory panel for consultation and regular feedback.

Figure 10.1: (*continued*)

- ✓ Create an online Millennial style guide that is updated regularly for covering Millennials in general and Millennials of color, women, and voters in particular. Include examples of best and worst practices for stories, headlines, photos and videos, social media, placement, sentences, quotes, word choice, tone, etc.
- ✓ Enhance smartphone and tablet computer news and apps to engage Millennials.
- ✓ Connect with Millennials through social media.
- ✓ Add introductory comments or prompts to stories posted on Facebook that will encourage Millennial engagement with news.
- ✓ News organizations that have made charging for news on the Internet part of their business model should waive the subscription fee for Millennials at least through age 26. (Poindexter, 2012, pp. 122–123)

Because Millennials engage with social media on their smartphones, news organizations can also reach Millennials by posting stories on social media, that is, if Millennials have "liked" their Facebook page or "followed" them on Twitter. The "Best Practices" in Figure 10.1 should also be applied to the stories that are posted and tweeted on social media.

While Millennials may be the social media generation, they are not the only generation getting news on social media, which may be one of the reasons some news organizations are expanding their presence on social media beyond a Facebook page or Twitter profile. Plus, some news organizations are participating on Snapchat's Discover, which has the potential to reach a younger mobile consumer. Snapchat is the latest addition to the social media landscape but so far it's not the gateway to news on mobile devices, according to a 2015 supplemental survey to the News for a Mobile-First Consumer Survey. In fact, according to Figure 10.2, browsers (30%), followed by apps (28%) are the most popular gateways to news on smartphones. As far as social media gateways, not surprisingly Facebook (18%) is the leader, followed by Twitter (9%). At 0.4%, Snapchat barely registers as a gateway to news.

The majority of the popular social media and photo and Video-sharing sites can be accessed by browser, search engine or app, but as an app, Snapchat must be downloaded to access it. As the newest social media platform, Snapchat is also the least downloaded app according to results of a supplemental survey to the original News for a Mobile-First Consumer Survey. When survey participants were asked which social media apps they have, results showed the following percentages: Facebook (73%), YouTube (67%), Twitter (41%), Google+ (34%), Instagram (32%), Snapchat (21%).

Just because Snapchat is not where consumers go for news does not mean news is not on Snapchat where it can be seen. But according to survey

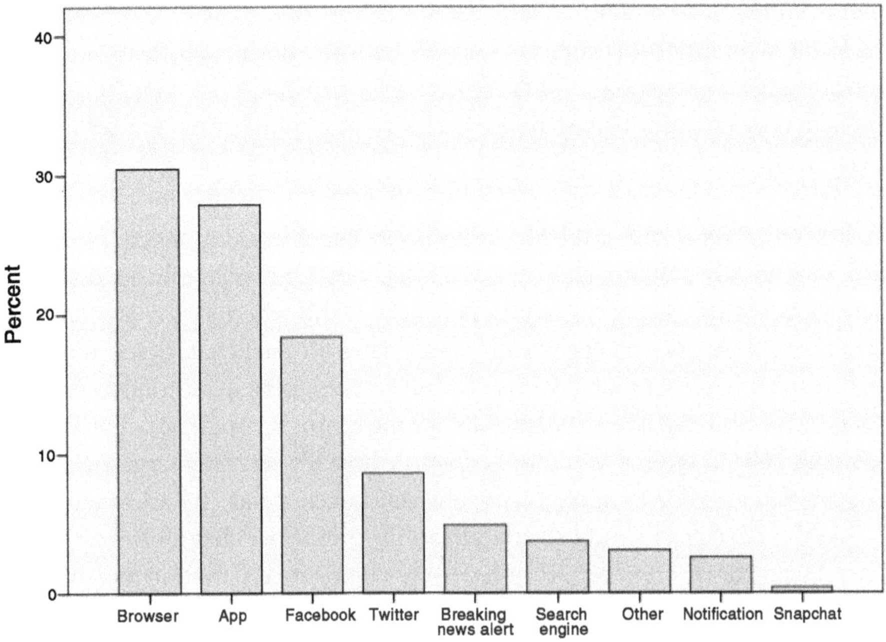

Figure 10.2: How Snapchat Compares as a Gateway to Mobile News.

participants, the overwhelming majority (71%) have never even seen news on Snapchat. In contrast, according to Table 10.2, which lists six social media and photo and video-sharing sites, it is rare for news *never* to be seen on Facebook and Twitter. Only 8% of Facebook users and 12% of Twitter users have *never* seen news on these social media sites. Although according to a *USA Today* article, Instagram has 400 million active monthly users, with one-quarter living in the U.S. (Guynn, 2015), the photo and video-sharing site is not a platform for news. According to Table 10.2, 62% of survey participants have "never" seen news on Instagram.

Table 10.2: Frequency That News Is Seen on Social Media Apps.

	Often	Sometimes	Occasionally	Never
Facebook	41%	25%	25%	8%
Twitter	35%	23%	29%	12%
YouTube	10%	17%	35%	39%
Google+	9%	12%	30%	50%
Instagram	6%	9%	23%	62%
Snapchat	5%	8%	16%	71%

And even for those who have seen news on Snapchat, only 4% say they have paid a lot of attention to the news, with 74% saying they paid little or no attention to news on Snapchat. While Snapchat is younger and it may be early in the adoption process, it must be underscored that taking your eye off proven social media to chase the next social networking site may not be the best strategy, because some social media sites may have limited value for posting and sharing news, especially if the site is primarily being used for another purpose. For example, when Snapchat users were asked what they primarily did on Snapchat, none of the responses mentioned news. And when survey participants were asked their opinion specifically about news on Snapchat, opinion leaned slightly more negative than positive, but one survey participant tried to balance the positive and negative: "Seems like it's meant for an audience that doesn't consume news as often as me. But I could see how it's very well targeted toward 17-year-olds who don't communicate any way but snapchat. It's adequate."

As the 2016 presidential election heats up, more Snapchat users may notice news. Snapchat is providing election news to engage its 100 million mostly teenage and Millennial users. Users can both shoot and upload election video and watch video snippets from the presidential campaign trail (Corasaniti, 2016).

Since Millennials are now the largest living generation and social media is here to stay and will always compete for attention on a mobile screen, Millennials and social media must be a part of any news organization's strategy to reverse declining news engagement. But if news is to be valued in the mobile world we now live in, the strategy to achieve that goal cannot be limited to news organizations even though news organizations must play a leading role because they produce the news product available on smartphones and tablets.

Essentials of Mobile Journalism

A plan for journalism and news to thrive in the mobile age, which I call "Essentials of Mobile Journalism," is outlined in Figure 10.3. This plan, which is comprised of eight essentials, is based on data discussed throughout *News for a Mobile-First Consumer*; past and current criticisms of journalism; the press's responsibilities and society's expectations as articulated by two national commissions, the *Elements of Journalism* and the Society of Professional Journalists' Code of Ethics; and my own experiences, which include working in the media industry on the business and news sides, researching news consumers and non-news consumers in a pre-mobile and mobile world, teaching Millennial journalism students, and creating local and national initiatives to increase news engagement and news audience knowledge. While each of the eight essentials in Figure 10.3 impact mobile journalism, the success of the overall plan begins with understanding the mobile audience and news consumer.

Figure 10.3: Essentials of Mobile Journalism: A Plan for Journalism and News to Thrive in the Mobile Age.

1. Understand mobile consumers and audience for news.
2. Develop *and* adopt guidelines for credible and engaging mobile news that consumers should know, news of interest to them, and news relevant to their daily lives.
3. Represent constituent groups—racial and ethnic groups, women, Millennials—in news coverage without stereotypes and in proportion.
4. Increase public understanding of journalism, its process and principles, and strengthen the public's trust in journalism.
5. Strengthen belief that in a democratic society everyone has a responsibility to be informed.
6. Cultivate news consumers among Millennials and the generations that follow.
7. Give mobile consumers and advertisers a reason to pay for news delivered on smart mobile devices.
8. Prepare for what's next.

Mobile Journalism Essential 1: Understand Mobile Consumers and News Audience

In the legacy media past, news organizations could be successful and profitable without knowing much about their audience but in today's competitive mobile space, audience ignorance will not increase news consumers or profits. In the past, legacy newspapers had little or no competition and local TV news competed with at most three other TV stations. A glance at a smartphone screen reveals the fierce competition in the land of mobile. Local newspaper, TV, cable, and radio news apps compete with news apps from around the country and the world.

Plus, there are non-news apps for social media, games, magazines, books, shopping, communicating, and streaming TV, movies, and live sports. The number and variety of apps displayed on a mobile screen underscore why it's so important to understand your target audience as well as others competing for your target audience's time and attention. While some in the newsroom may take offense at the idea of thinking about the news as a product, albeit different from a box of cereal, bar of soap, or a car, news is still a product that is produced and delivered throughout the day, every day, to an audience that has the power to engage with it from a variety of platforms any time of the day or night or ignore it.

When reading a newspaper was not just an institutionalized behavior, but an everyday habit, and in some homes a ritual, the audience for newspaper news had a known profile—middle-aged, educated, higher income, more male than female, more white than non-white. And the newspaper reader was a loyal subscriber whose paper was delivered to the driveway or front door every day of the week before dawn.

But in today's mobile world delivering a newspaper to a subscriber's driveway before dawn seems quaint, and even going to a news website on a laptop or desktop has ceased to be the norm. In the mobile world, consumers may go to a news app or use their browser or a search engine to arrive at a mobile news site but they are equally as likely to forego the app, browser, and search engine and click on a story because of a breaking news alert, tweet, email notification, Facebook post, or a "share" by a friend, family member, or colleague. Because of the variety of ways that a consumer arrives at a story, understanding *who* is in the audience is not just complicated but may be impossible to truly know. Clicks from unique visitors don't tell anything about who looked at or read, a specific news story. At least TV news has ratings with some demographic information, even if they are not infallible. But whether the audience is more of a mystery than a definitive demographic, understanding the mobile audience is necessary if the declining news engagement trend is to be reversed and a profitable business model is to be developed. Therefore, all newsroom staff—not just a select few—need to understand the audience that both consumes and ignores the news. Knowledge of the audience should include mobile consumer types, generational characteristics, demographics, attitudes, beliefs, motivations, and mobile news consumption behavior.

News for a Mobile-First Consumer has laid the groundwork for understanding consumers in the mobile landscape by creating the first taxonomy that sorts mobile device owners according to how they primarily use their smartphones and tablets. This taxonomy should serve as the underpinning for discussions about mobile news consumers as well as establish a baseline that future changes can be measured against. Over time as more mobile device owners adopt a mobile-first or mobile specialist approach, fewer mobile laggards will exist. But in the meantime, wherever consumers may be in the adoption cycle, the news audience needs to be understood in the context of the three mobile consumer types.

The news audience also needs to be understood in the context of the generation in which it was born. Historically, research on news consumers by academics, news organizations, and the Pew Research Center has used age rather than generation to provide insight into news consumption behavior and attitudes. And while age has been useful in distinguishing news consumers from non-news consumers, as a variable, age does not automatically include the context and nuance of the factors influencing a generation based on the year it was born.

Millennials, like generations that preceded them and generations that will follow, do not exist in a vacuum; growing up, generations are first influenced by the generation their own parents belong to. The time period that a generation grows up in, as well as when they come of age into adulthood, are also influencing factors. While historical and economic factors may leave a permanent imprint on a generation, when it comes to developing into news consumers, a host of other factors, including media, may intervene. (Poindexter, 2012, p. 16)

It's these factors, which have influenced generations born in the 20th century but now living in a 21st century mobile world, that have made a person's generation the most influential factor in distinguishing news consumers and mobile consumer types. News consumers are more likely to be among Baby Boomers while mobile-first consumers and mobile specialists are more likely to be Millennials. The fact that the Millennial Generation is also the most racially and ethnically diverse and Baby Boomers are the least diverse helps explain why more Latinos and Asians Americans are among mobile-first and mobile specialist consumers but whites are more likely to be news consumers. Millennials are also less likely than Baby Boomers to believe being informed is important, yet the motivation to be informed about what's going on is one of the reasons for consuming news. Called "uses and gratifications" by journalism scholars, news consumption motivations, which were first studied in the 1940s, are also relevant almost seven decades later in today's mobile landscape (Berelson, 1949; Katz, et al., 1974). In fact, the desire to know what's going on around you, also called surveillance, drives news consumption behavior more than any other motivation. Some motivations such as entertainment and social reasons may better explain Millennial news consumption, while motivations for information and opinion provide insight into news consumption by older adults (Lee, 2013). Motivations that have been identified over the decades that are also relevant in the mobile landscape include diversion, passing time, tool for daily living, social prestige, and joy of being informed (Incollingo, 2015).

Understanding the news audience also means understanding their news preferences. As found in Chapter 4, national, local, and weather are not only among the most popular types of news consumed on a daily basis, they represent news that most motivates consumers' news engagement. Breaking news, which is also among the most popular types of frequently consumed news, has a special appeal for mobile-first and mobile specialist consumers because wherever they are when news breaks, they are just a finger tap away from knowing what is happening, when it is happening.

While some research describing news consumption behavior in a pre-mobile landscape is still relevant, it must be emphasized that *how* consumers engage with news on a smartphone is very different from reading a newspaper or watching news on TV: more solitary than communal, more interactive than one-way, and more like eating tapas spread out over the day than eating a full-course meal in one sitting. In testing his theory about how consumers engage with news on smartphones, journalism scholar and former newspaper editor Logan Molyneux (2015), who also conducted the first national study of journalists' use of smartphones as a reporting tool, found "smartphone news sessions were shorter and spread throughout the day" (p. 135). In fact, when accessing news on smartphones is compared with other platforms, the difference between newspapers and smartphones is striking. As shown in Table 10.3, 85% of newspaper readers consumed news from newspapers all at once, which was different from consumers who accessed news on smartphones. Almost three-quarters (72%) accessed news on smartphones throughout the day. Getting news on laptops and desktops was similar to getting news on smartphones but for accessing news on tablets and radio there was more of a balance between all at once and spread out throughout the day. Despite the difference in watching TV news versus reading a newspaper, getting news all at once from TV was more similar to how newspapers are read. In fact, TV news viewers were twice as likely to get news all at once rather than throughout the day.

Table 10.3: How Consumers Access News on Different Platforms.

	All at Once	Spread Out
Newspapers	85%	15%
TV	66%	34%
Radio	52%	48%
Tablet	47%	53%
Laptop/Desktop	29%	71%
Smartphone	28%	72%

Since mobile news consumers check their mobile devices for news throughout the day, it is imperative that they find fresh news. New stories can be posted and old stories can be updated with new information. And checking news spread across the day suggests news is being consumed in shorter periods, which also means long stories, even when compelling, may not be as effective on smartphones. That's why how news is read by the three mobile consumer types must be considered in developing the most effective format to engage smartphone consumers. Mobile specialists who are news seekers will likely read every screen of a long story while mobile specialists who are social media devotees or gamers or even mobile-first consumers or mobile laggards may leave a story after reading the first screen or two.

Although much more research needs to be done, one thing is clear: All mobile consumers are not the same and any strategy to increase news consumption should not assume a one-type mobile news format fits all.

Mobile Journalism Essential 2: Guidelines to Mobile News Excellence

Developing *and* adopting guidelines for mobile storytelling to produce engaging news reports with the highest journalistic standards is not unrelated to understanding mobile news consumers. By thoroughly understanding who the audience is and how the audience engages with news on smartphones, journalists can better deliver the quality of journalism that the public has a right to expect. Furthermore, by incorporating audience-engaging techniques recommended by journalism scholar Jake Batsell (2015) in his recently published book, *Engaged Journalism: Connecting with Digitally Empowered News Audiences*, the news consumption experience can be enhanced, which can ultimately strengthen the public's belief that being informed is important.

When journalism fails on mobile, the obvious culprit is that, as a platform for news, smartphones have limitations that can unintentionally mislead, distort, or even bore. Since news reports written for non-smartphone platforms do not easily transfer to a smartphone screen, newsrooms have to be vigilant about making sure the small screen format is an accurate representation of the original story and not a sub-standard version because the headline is less crisp, or the anecdotal lead took so many screens to develop that the reader got bored and moved on, or the multimedia that was amazing on a laptop was a bust on a small screen.

So many things can go wrong that can make mobile news less than the best from the mobile consumer's perspective that newsrooms, especially those grounded in legacy newspaper routines, must speed up the adoption of mobile-first newsroom routines so the news that consumers engage with on their smartphones will be as engaging as news posted on non-mobile platforms.

What would a news story look like if it were written for mobile-first? Would it be inverted pyramid or feature? Or BuzzFeed-type listicals? When a BuzzFeed story treatment was tested against a *USA Today* story treatment in an experiment with college students, most participants preferred the BuzzFeed style, but they retained less information than the *USA Today* approach to the story (Burton & Roberts, 2015).

Would some writing styles, such as an anecdotal lead, be banned from smartphone reports? Should there always be at least two versions of a smartphone story? Original story plus summary? And what about a tweet-length account of the story? Should the day's mobile news always be introduced with the editor's picks for "need to know" and "don't want to miss" stories? And how should breaking news be treated? Always introduced with a breaking news banner? And what about updates? And because of the public's perception that the news media are unwilling to correct mistakes, how should mobile corrections be handled? And should there be written mobile-first news guidelines similar to the social media guidelines in *The Associated Press Stylebook*?

These questions have grown out of a three-year, one-person "experiment" that I have been conducting with my mobile device to understand firsthand how consumers engage with news on their smartphones. I started the experiment by canceling my print subscriptions and only reading newspapers on my phone. Because I was so determined to understand what mobile-only news engagement felt like I wouldn't even read free printed newspapers delivered to my hotel room door when I traveled. I literally have not read a printed newspaper since May 2013.

As someone who loved reading two print newspapers every day, I initially missed my daily ritual, but it didn't take long for me to enthusiastically embrace my new mobile news routine. While I technically am not mobile news-only since I do watch network TV news, cable TV news, and some local TV news, I am mobile news-first because I get news from my iPhone in the morning and don't watch TV news until the evening.

In my mobile news experiment, I discovered that I loved news apps, which initially had me wrongly assuming that others loved and used news apps as much as I did. A business executive who sat next to me on the plane told me he was very frustrated reading news on his smartphone, which I once again wrongly assumed that he was talking about apps. Trying to better understand his frustration, I asked several questions and learned the frustration wasn't coming from using apps because he wasn't even aware that news apps existed. My Millennial students are aware of news apps but they don't download them because they take up too much space on their phones. The lesson for me was to ask questions and not assume in this new mobile environment, because not only is there a lot to learn, we don't even know everything we don't know.

Using only my smartphone for news, I'm also keenly aware of the competition on a mobile screen for a consumer's attention. On a mobile screen, newspapers compete directly with other newspapers, as well as with network and local TV news, which compete with cable and radio news. And while the news on a mobile screen has distinct brands, the news product produced by legacy newspapers and broadcast is very similar—text and photos, supplemented with multimedia, and most recently virtual reality. Of course, the competition on a mobile screen does not end with news because social media, games, entertainment, streaming, and other types of apps also compete for consumers' time and attention.

My mobile-first news engagement routine begins every morning when I tap the apps and read news from CNN, the *Washington Post*, NPR, the *New York Times*, the *Los Angeles Times*, *USA Today*, BBC, and NBC Nightly News. Several times a week I also check BuzzFeed and Huffington Post. This routine of tapping, scrolling, swiping, and reading news stories, from top stories to world, national, business, technology, and science and health has enabled me to see news engagement through the eyes of a mobile news consumer. What I've learned is that all storytelling formats are not effective on mobile and when the least effective story format is used on mobile, the story does not engage and may even repel. How stories look on mobile screens is even more problematic when it comes to race and ethnicity, women, and Millennials.

Mobile Journalism Essential 3: Represent Constituent Groups

While all story formats are not effective on mobile, the small screen vertical layout can unintentionally distort the news, a problem made worse when covering racial and ethnic groups, women, and Millennials. These constituent groups may be different, but there are similarities in the media's treatment—exclusion, stereotyping, and failure to cover in proportion—that is exacerbated on a small mobile screen.

In 1947 when the Hutchins Commission Report stated that projecting a representative picture of the constituent groups in society is one of five requirements the press is expected to meet, the U.S. population was significantly less diverse. By 1965, the U.S. population was still overwhelmingly white, with only 15% of the population racial or ethnic minorities (Linthicum, 2015). Today 37.2% of the U.S. population is Latino, African American, Asian American, and Native American and 50.8% are women, but news coverage remains overwhelmingly white and male. Millennials are now the largest living generation, but they are rarely included in the news as subjects or sources.

Rather than fairly and accurately representing people of color, the news media are more likely to stereotype them. In fact, research has found the news media have stereotyped African Americans, Latinos, Asian Americans, and Native Americans in 18 different ways: outsiders, historic and cultural relics, exotic, noble savages, degraded, bad and uncivilized, vanishing race, poor, criminals, threatening, confrontational, problems, society and government critics, athletes, entertainers and celebrities, gang members, model minority, and emotional (Poindexter, 2011).

Research has further found that the news underrepresents racial and ethnic groups as crime victims; in everyday activities; as achievers; and as official, expert, and private citizen sources (Poindexter, 2011). The stereotypes of racial and ethnic groups appear to have been transferred from legacy and laptop news platforms to smartphones, where small screens exaggerate stereotypical images. For example, the news media include African Americans in news stories about poverty and crime and report on them as entertainers and athletes, but those images do not represent the more than 42 million African Americans in the U.S. Research has shown that the news covers Latinos and Native Americans, if they are covered at all, as "problem people" and "despite what appears to be mutually exclusive labels, the news media stereotype Asian Americans as both gang members and model minorities" (Poindexter, 2009, p. 437).

While research has shown that historically Latinos have been excluded from news coverage, during the 2016 Republican presidential primary, the news media, by failing to provide context, appear to have placed all Latinos, whether U.S. citizens or not, under one umbrella that some Republicans have used to debate illegal immigration and oppose a pathway to citizenship for undocumented immigrants. In other words, the news media have continued the practice of reporting on Latinos as "problem people," but they have failed to report the immigration debate with clarity and context, which has allowed for the emergence of a new stereotype of Latinos as undocumented immigrants.

Similar to racial and ethnic groups, women have been excluded from the news and not used as official and expert sources. Studies of the front page of a newspaper, which was once used as a reliable way to assess news coverage of women, found women and issues of interest to women were less likely to be treated as newsworthy subjects on the front page and were relegated to "second-tier" status on the inside or feature pages that were once called women's pages (Poindexter & Harp, 2008). Because there is no front page on mobile devices in the traditional newspaper front page sense and the top news story changes throughout the day, there is no equivalent reliable measure to assess coverage of women or issues of interest to women. Still, scrolling down a smartphone screen of top stories or swiping through the first story of each content section finds few stories about women even though they represent 51% of the U.S. population.

The news media's biased coverage of women who have aspired to the highest office of the U.S. has been well documented (Cáceres, 2011; Poindexter, 2008b). Quotes of women candidates have been paraphrased while men have been quoted directly. The news media have emphasized the appearance of women, including their attire, while rarely reporting on the appearance of a male candidate for president. The news media's focus on the appearance of female presidential candidates has been such a part of the press's presidential news coverage that at the 2015 White House Correspondents' Dinner, *Saturday Night Live*'s Cecily Strong, the MC at the exclusive journalism event, made the journalists attending the dinner promise not to cover Hillary Clinton's appearance during the 2016 presidential campaign since that wasn't journalism (Dittmar, 2015).

A comparison of words used to describe Hillary Clinton versus Barack Obama during the 2008 Democratic presidential primary election found explicit bias. Hillary Clinton was described by the media as "ruthless," "unsympathetic," and "utterly cold" while Barack Obama was described by the media as "kind," "sympathetic," and "emotionally accessible" (Cáceres, 2011, p. 122). With Hillary Clinton the only woman running for the Democratic nomination in the 2016 presidential election and Carly Fiorina the only woman running for the Republican nomination, scholars no doubt will analyze 2016 presidential election news to determine if news coverage of female candidates improved, worsened, or remained the same.

While women have been relegated to "second-tier" status if included in the news at all, and women presidential candidates have been covered in ways that men would never be covered, Millennials are rarely included in the news unless someone in their generation commits a crime, is in an accident, or achieves celebrity status (Poindexter, 2012). In the pre-mobile news age, journalists stereotyped Millennials as selfish and gleefully mocked them for having to move home after college graduation or being weighed down by student loan debt. In the mobile news age, Millennial mocking has for the most part disappeared from the news, but it hasn't been replaced by including Millennials as newsworthy subjects or sources.

Developing and adopting a plan in which racial and ethnic groups, women, and Millennials are included in mobile news without stereotypes and in proportion is not just what is expected of journalism; it's good business and may be the only way to reverse declining news engagement that is particularly acute among Millennials, women, and racial and ethnic groups. Until developing and adopting a plan to include Millennials, women, and racial and ethnic groups in mobile news coverage is deemed a newsroom priority, mobile news will continue to be a product that falls short of what the public expects and journalism's responsibility.

Mobile Journalism Essential 4: Increase Public Understanding and Trust

If the public better understood the process and principles of journalism, would the public be able to distinguish between good and bad journalism? Would more public understanding about journalism mean more trust? And what, if any, impact would more transparency about press operations have on news engagement and the public's perception of the role of the press in a democratic society?

Because the mobile landscape is populated with professional journalists; citizen journalists; partisans; unverified tweets and posts as well as technology and social media companies that aggregate news, the public would benefit from learning the standards that journalism aspires to operate by. Knowing journalism's principles would give the public standards by which to evaluate news reports and other types of information for their credibility. Plus, this education would signal that journalism's goal is transparency—not secrecy—which could elevate journalism's credibility in the eyes of many and, perhaps, lead to increased news consumption and a willingness to pay for news on mobile devices because its value is better understood.

Since there are no educational, exam, or licensing requirements to be a journalist in the U.S., technically anyone can be a journalist. But from the vantage point of Tom Kent, the standards editor for the Associated Press, who participated on the 2014 AEJMC presidential panel (AEJMC, 2014a) that graded the press and journalism schools, the only people who can claim to be journalists are those who occupy themselves "full-time with gathering and transmitting information" using "fact-based judgments" and are guided by a "visible transparent code of ethics."

Therefore, to elevate the public's trust in journalists and journalism, it is imperative that the public understand who is a journalist as well as the process and principles of journalism, with specific attention paid to verification and why it distinguishes journalism from stuff that tries to pass as journalism. Verification, according to the authors of the *Elements of Journalism*, is guided by the following core concepts:

1. Never add anything that was not there originally.
2. Never deceive the audience.
3. Be as transparent as possible about your methods and motives.
4. Rely on your own original reporting.
5. Exercise humility. (Kovach & Rosenstiel, 2014, pp. 111–112)

The first four core concepts are self-explanatory, unlike "exercise humility," which is to me a nice way of saying avoid being arrogant, assuming you know everything, because you may not know what you don't know, which makes verification difficult if not impossible. These core verification steps cannot guarantee that mistakes will not be made in journalism, but they show a commitment to journalism's highest standards and a level of transparency that can only elevate the public's trust in the press and its contributions to the public good.

Mobile Journalism Essential 5: Strengthen Responsibility to Be Informed

That there is a responsibility to be informed is not just a nice idea or an untested theory. Table 10.4 shows there is a correlation between believing we have a responsibility to be informed of current events and actual news engagement. Participants in the News for a Mobile-First Consumer Survey who strongly agreed they had a responsibility to be informed about news were significantly more likely than survey participants who strongly disagreed they had a responsibility to be informed, get news seven days a week (74% vs. 33%). A similar pattern was found for seeking news; those who strongly agreed we have a responsibility to be informed were two and a half times more likely than those who strongly disagreed with the statement, seek news seven days a week (60% vs. 24%).

Although correlation is not causation, the fact that this relationship was first documented in the pre-Internet landscape, was replicated in the Internet news landscape, and has emerged in the mobile landscape is a testament to the strength of the idea of the responsibility to be informed. This responsibility does not suddenly appear from nowhere. As discussed earlier, believing that it is important to be informed and news consumption are primarily the product of socialization in the home and schools. Just as someone can be raised to believe being informed is important, one can be raised to believe it is a waste of time to be informed of the news.

Table 10.4: Relationship between Responsibility to Be Informed and News Engagement.

	Strongly Disagree	Somewhat Disagree	Neither Agree Nor Disagree	Somewhat Agree	Strongly Agree
Get news 7 days	33%	33%	42%	54%	74%
Seek news 7 days	24%	20%	27%	40%	60%

Mobile Journalism Essential 6: Cultivate Millennials and Generations That Follow

Other than programs that made discounted newspapers available to schools and college campuses, news organizations have not invested a lot in developing future news consumers, and they have done even less to promote the idea that, in a democratic society, being informed is an individual and public-good benefit. That lack of investment in future news consumers cannot continue in today's mobile landscape; there is simply too much at stake for individuals, the public at large, our democratic society, and news organizations in need of a sustainable business model. Perhaps, more people would be willing to pay for a digital subscription if they believed in the individual and public good benefits of being informed.

Four years ago before we knew the extent of the mobile transformation of the news landscape, I proposed a detailed strategy for cultivating news consumers among Millennials (Poindexter, 2012). Based on the results of the survey I conducted for *Millennials, News, and Social Media*, I recognized that Millennials were different from previous generations, which is why I became concerned that unless Millennials were cultivated to become news consumers, news engagement could become a thing of the past. Because mobile devices have made Millennials who are mobile-first and mobile specialists even more different from previous generations that are mobile laggards, cultivating them as news consumers is more urgent, and the following four recommendations are even more relevant in today's mobile landscape:

- ✓ Improve news coverage of Millennials.
- ✓ Legacy news media must get involved in increasing news engagement.
- ✓ Social media must play a role in facilitating news engagement.
- ✓ Establish a national day for news engagement.

There's no question that news coverage of Millennials has improved since 2012 when I first made this recommendation in *Millennials, News, and Social Media: Is News Engagement a Thing of the Past?* And although newspaper-owned, young adult–targeted publications produced by legacy newspapers in San Antonio and Dallas no longer exist and Chicago's *Red Eye* may have more non-news, including sponsored content, than news, there are more than a half dozen Millennial-appealing news platforms that can be accessed by Millennials on their smart mobile devices (see Table 5.5). While these news platforms represent different approaches to the same goal of encouraging and increasing Millennial news engagement, they alone are not enough. Just as there is a sustained commitment to journalism of the highest standards, there also must be a sustained commitment to improving news coverage of Millennials. Poor coverage of Millennials only lowers journalism's high standards because Millennials are one of society's constituent groups that must be accurately represented in the news.

Legacy news organizations must also get involved and stay involved in increasing news engagement among Millennials. The *New York Times* with its NYT Now news app, which it made free one year after initially charging a monthly subscription fee, is leading the way, and now the American Press Institute (API), a major resource for legacy media, is conducting research on Millennials and recommending strategies for journalists and publishers to reach out to Millennials (Maerz, 2015). It is hoped that legacy media's newfound interest in Millennials is a sustained commitment to what is now the largest living generation and not a short-term commitment as it has traditionally been in the past. In fact, the authors of the 2007 edition of *The Elements of Journalism* admitted that news organizations "did nothing to help create a new generation interested in news" (Kovach & Rosenstiel, 2007, p. 211).

In 2012, I also recommended that social media play a proactive role in facilitating news engagement, but what we now know about the mobile landscape is that social media sites are competing with news sites for the time and attention of mobile consumers and so far social media apps are winning. Because for Millennials, Facebook and Twitter are the pathways to news, news organizations have to balance cultivating Millennial news consumers through social media and mobile devices and building a bridge to a business model that takes into consideration the needs and interests of Millennials while cultivating them as news consumers and future digital subscribers.

My recommendation in 2012 to establish a national day for news engagement grew out of my long-term concern about declining news consumption and my goal of once again becoming a nation where being informed was valued. Below is how I envisioned this day set aside for news engagement:

> Support a national "Engage with News" Day to be held the first Tuesday of October. Teachers would bring news into the classroom; families would read, watch and discuss news at home together. Journalists, representing their news organizations and journalism associations, would visit classrooms to emphasize the purpose, principles, and process of journalism. The news media and journalism classes would cover the "Engage with News" Day event and these stories with accompanying multimedia and lots of quotes from Millennials could be posted and shared on social media. (Poindexter, 2012, p. 131)

During my term as president of the Association for Education in Journalism and Mass Communication (AEJMC), I had an opportunity to make my vision of News Engagement Day a reality on the first Tuesday in October 2014 (Poindexter, 2014). Under the sponsorship of AEJMC, journalism and communication faculty and students, journalism and communication professionals and associations, public and private K–12 schools across almost all 50 states and the District of Columbia, as well as other countries, participated in this day-long event "to revitalize the public's engagement with news, regardless of generation" and be reminded that being informed of news is empowering, enjoyable, and "essential to our democratic nation's future vitality" (Poindexter, 2012, p. 132).

Since my 2012 News Engagement Day recommendation, the news landscape has undergone a transformation where Facebook with Trending, Twitter with Moments, Snapchat with Discover, and the iPhone with its Apple News app have joined legacy news media, digital-only news media, and other news organizations to provide mobile news. And whether we're in today's mobile landscape or have transformed into what comes next, engaging with news and being informed are vital to a democratic society, and News Engagement Day provides us with an opportunity to reinforce that message on the first Tuesday of every October.

Mobile Journalism Essential 7: Make Stronger Case for Mobile Subscriptions and Ads

Table 10.5 shows that survey participants in the News for a Mobile-First Consumer Survey who strongly agree that they have a responsibility to be informed are significantly more likely to have a news subscription than those who don't believe this (34% vs. 0%). But there is a catch. These same participants who strongly agree that they have a responsibility to be informed *also* strongly believe news on the Internet should be free. Almost 7 out of 10 (68%) who strongly agree that we have a responsibility to be informed about news and current events also strongly agree that news on the Internet should be free.

Table 10.5: Relationship between Responsibility to be Informed and Paying for News Subscription.

	Strongly Disagree Responsibility to be Informed	Somewhat Disagree	Neither Agree Nor Disagree	Somewhat Agree	Strongly Agree Responsibility to be Informed
Pay for news subscription	0%	16%	19%	26%	34%
Do not pay for news subscription	100%	84%	81%	74%	66%

And, perhaps, of most concern is Table 10.6, which compares generations and having a news subscription. When asked, "Do you pay for any news subscriptions? the majority of every generation responded, "No." Furthermore, as generations get younger a larger percentage do not have a news subscription. Just over half (52%) of Pre-Baby Boomers do not have a subscription to any type of news but almost 9 out of 10 (86%) of the youngest Wave II Millennials do not have a subscription to news.

Table 10.6: Generations and News Subscription.

	Pre-Baby Boomer	Baby Boomer	Generation X	Millennials Wave I	Millennials Wave II
Have news subscription	48%	34%	23%	17%	14%
Do not have news subscription	52%	66%	77%	83%	86%

To provide further insight into paying for a subscription by Millennials, Table 10.7 re-analyzes digital subscription data from API's Millennial survey. Of 14 digital subscriptions asked about, Millennials were least likely to pay for news-related subscriptions. While 45% paid for subscriptions to rent or stream movies, only 12% paid for a print newspaper subscription, 11% paid for digital news apps, and 7% paid for a digital newspaper subscription.

Table 10.7: Millennials Paying for Subscriptions, Products, and Services.

Movies or TV shows rented or streamed through iTunes, Netflix, etc.	45%
Music downloaded or streamed	39%
Video games or gaming apps	38%
Cable TV	35%
Other apps	29%
Print magazines	16%
E or audiobooks	16%
Online course	15%
Print newspapers	12%
Digital news apps	11%
Digital subscriptions for magazines	8%
Digital subscriptions for newspapers	7%
Email newsletter	6%
Sample box service such as Birchbox or Goodebox	6%

Note: Compiled from How Millennials Get News (2015); Topline Questions, 37–40.

This belief that news on the Internet should be free and the willingness of Millennials to pay for anything *not* news may be why one executive described the failure to monetize the "unfettered growth in mobile" as an "Oh Sh*t Mobile Moment" (Sebastian, 2015b). What then, if anything, can be done to persuade more consumers, whether Millennials, Generation X, or Baby Boomers, to pay for mobile news and more advertisers, whether legacy media or digital-only, to buy mobile news ads? Adoption theory may have some answers.

When a consumer decides to subscribe to news that's delivered on a mobile device or an advertiser decides to buy ads on a news app or mobile news site, that is adopting a new process or behavior. Since, as Everett Rogers (2003) theorized, adopting a new idea, device, process, or behavior is a function of relative advantage, compatibility, complexity, trialability, and observability, news organizations can develop and test marketing strategies that emphasize these characteristics when trying to persuade consumers to adopt mobile news subscriptions.

For example, since the opportunity for trialability can contribute to adoption of an innovation, news organizations can develop a marketing strategy in which consumers are offered a three-month free subscription with the goal of converting the free trial subscription into a paid subscription once the free subscription expires. That marketing approach will likely produce more paid mobile news subscribers than the 10 free articles a month that some news organizations use. Plus, the three-month free trial has the potential to build loyalty to both the brand and the brand experience.

In terms of building loyalty, it cannot be emphasized enough that there's a big difference between experiencing a mobile news brand as a subscriber versus randomly reading articles until you've maxed out for the month. No doubt that's why Apple offered a three-month free trial when it launched its Apple music streaming service, and before Christmas, the *New York Times* invited valued digital subscribers to give a three-month digital gift subscription with unlimited access to NYTimes.com and NYT apps to two friends or loved ones who were not digital subscribers. When I received the *New York Times* email offer, I immediately sent the email addresses of my sister and my daughter to the *New York Times*. In less than 24 hours, I received thank you text messages from my sister and daughter saying they were thrilled to receive the three-month digital news subscription gift.

Two of the reasons this marketing strategy to get non-subscribers to adopt digital subscriptions is so clever is that it made me, the subscriber, feel good and I was able to do something that the *New York Times* could not do easily or cheaply—identify the best prospects for a subscription to news delivered on a smartphone, tablet, or laptop.

Ever since the newspaper business became less profitable in the digital landscape, news organizations have been trying to figure out how to replace that lost subscription and advertising revenue. In fact, as recently as October 2015, the Wharton School of Business sponsored an invitation-only, off-the-record summit for what it called "c-suite decision-makers" who paid $3,000 each to assess and address revenue streams that may deliver the elusive profits that newspapers effortlessly produced in a pre-digital world (The media future summit, 2015). Ironically, the people closest to the problem who may have had the best ideas for making current revenue streams more profitable and identifying new revenue streams whether from consumers or advertisers were excluded from the event.

In an industry that *USA Today* columnist Rem Rieder (2015b) described as still searching for the "right formula" to "survive and thrive in the digital era," everyone's voice needs to be heard. And in the mobile world that we live in today, the best ideas for news organizations to survive and thrive in the mobile era could come from news organization employees who are closest to the problem, as well as advertisers, journalism scholars, or even mobile-first consumers, mobile specialists, and mobile laggards.

Mobile Journalism Essential 8: Prepare for What's Next

In 2009 when a journalist-in-residence at the University of Texas at Austin School of Journalism showed off his iPhone to journalism students during a panel focused on challenges facing the future of journalism, no one could have predicted that the news landscape would have been transformed to the mobile landscape of today. And who could have predicted that three distinct types of mobile consumers—mobile-first, mobile specialists, mobile laggards—would emerge? Preparing for what's next starts with understanding and tracking these three types of mobile consumers.

While it's impossible to know exactly how long it will take for mobile-first consumers to dominate the mobile landscape, it is knowable that news organizations should prepare for that eventuality by developing and adopting strategies that will make news a priority in the lives of mobile-first consumers. Similarly, news organizations must develop strategies to increase news seekers among mobile specialists who currently represent only 6% of that mobile consumer category.

News organizations cannot afford to ignore the factors identified in *News for a Mobile-First Consumer* that correlate with higher levels of news consumption. If, for example, users of news apps engage more with news, then news organizations should promote their news apps, especially among those who are less interested in news. Or if higher levels of trust in journalism correlate with more news engagement, then news organizations should do more to make the process of doing journalism more transparent in order to build more trust.

Preparing for what's next also requires long-term thinking and research about the post-Millennial generation so that news will have a place on their mobile screens and in their lives. News organizations have an opportunity for a do-over with this generation, but it will be a challenge to cultivate news consumers among the generation born between 2000 and the end of the second decade of the 21st century. This do-over requires a long-term commitment and resources dedicated to creating news consumers in the post-Millennial Generation in which some are starting high school, some are infants and toddlers, and others have yet to be born. While Millennials could be called the social media generation, post-Millennials could be labeled the Mobile-First Generation. Two-thirds (67%) of teens and one-quarter (24%) of tweens in the post-Millennial Generation own smartphones and they are using them every day for virtually everything (The Common Sense, 2015). Excluding texting and talking on the phone, tweens are averaging over three and a half hours on their smartphones and teens are averaging over four and a half hours every day.

A study published in the journal *Pediatrics* found post-Millennial teens and tweens aren't the only ones in this generation engaging with mobile devices. The youngest of the post-Millennial Generation are engaged with smart mobile devices before they can "talk and walk" (Maltais, 2015). In fact, 97% of the children ages 6 months to 4 years had used a mobile device and most of the 3- and 4-year-olds could navigate the mobile devices without help.

As I watched two-and-a-half-year-old Harper, who lives in Kansas, navigate a smartphone, it occurred to me that this post-Millennial is not just navigating a mobile device as the *Pediatrics* journal study found, but mobile devices may very well be in the DNA of this generation, which is why mobile-first captures what makes this generation unique, unlike the nondescript Generation Z label that some journalists and marketers are using to brand this generation.

Harper effortlessly taps and swipes Fisher-Price's "Laugh&Learn" app that uses sounds, color, animation, and music to teach colors and shapes. With help in spelling kangaroo, horse, elephant, clown, and cat on the Google app,

Harper Googles then swipes though the images that appear on the screen. By Googling animal, rabbits, lions, zebras, pandas, and a cornucopia of other animals pop up on the screen, leading to intense swiping and giggling from the two-and-a-half-year-old. Harper also plays and dances to downloaded and streaming music on the smartphone and she's not just playing children's songs. She's equally adept at finding and playing a family video or *Nick Jr.* on her mother's iPad mini and she's a regular on FaceTime where she talks to her grandmothers and shows off her books, toys, and puzzles.

Just like her mother, Harper is growing up with news around her but it's less obvious because Harper's mother primarily gets news on her smartphone. That will be one of the challenges of growing news consumers among the post-Millennial Generation: socializing news engagement by example when news is swiped, tapped, streamed, read, and shared on a smartphone is just not as effective.

That's why creating news consumers in the post-Millennial Generation requires including parents and teachers in any strategy because they have the most influence, that is, until media, social media, and peers begin to replace parental and teacher influence in the pre-teen years. News Engagement Day is one vehicle that news organizations can adopt to reach both parents and teachers to encourage news engagement. Historically NIE (Newspaper in Education), a legacy newspaper program created to cultivate future news consumers through the schools, could also be very effective but it is in need of more resources and committed leadership (About NAA, n.d.). When NIE was a priority in the newspaper industry, more than 950 newspapers delivered newspapers and educational materials to 40% of public school students. According to the results of my survey of NIE managers, 84% of the educational materials emphasized the "importance of being informed," "78% put the focus on the First Amendment," and 60% of the programs emphasized the "purpose and principles of journalism" (Poindexter, 2012, p. 96).

The public's low regard for the press may also be a function of lack of awareness of the importance of the press to the First Amendment to the U.S. Constitution. In fact, the AEJMC Poll on Public Attitudes Toward the Press reported News Engagement Day 2015, found that only 22% of the public could correctly name the press as one of five freedoms in the First Amendment. Additionally, 95% of the public could not name all five freedoms, and 15% incorrectly named the Second Amendment right to bear arms as a First Amendment freedom (Can Americans name the five freedoms, n.d.).

Despite the decline in the priority of NIE, some newspapers remain active in the schools, providing digital versions of the news and curriculum materials. And the *New York Times* continues its long-time commitment to providing news and educational materials to schools nationwide. In fact, "in honor" of the 2014 inaugural News Engagement Day, the *New York Times* Learning Network posted "50 Ways to Teach with Current Events" (Gonchar, 2014).

While almost everyone is a mobile consumer, even teens, tweens, and toddlers, not everyone is a *mobile news consumer*. And for some mobile consumers, news has no place on their mobile screens or in their mobile lives. If news has no place on a mobile screen or in the lives of mobile consumers, it's impossible to make a persuasive case for paying for mobile news. And unless the case for paying for news on mobile devices is made to consumers and advertisers, news organizations will not have the financial strength to gather and report credible news the public wants and needs.

The catch-22 situation that news organizations find themselves in is partly due to the fact that historically news organizations have been more like laggards than innovators in the adoption cycle. News organizations have too often stubbornly held on to the old ways of doing things even when their news landscape was changing around them. *News for a Mobile-First Consumer* confirms that it is not just a transformed news landscape out there; a completely different type of consumer now inhabits the news landscape. And news organizations that haven't already adopted a mobile-first culture will fall further and further behind because wearable news is already here in the Apple Watch, virtual reality, which will provide innovative ways to report the news, has also arrived, and yet-to-be-imagined innovations aren't far behind (NYT VR, 2015).

But preparing for what's next is not just about technological innovations. Preparing for what's next also means keeping up with the pulse of consumers, including the generation that will follow Millennials, and creating news consumers among them. To both keep tabs on the pulse of consumers and create new, news consumers requires insight beyond industry analytics and assumptions that all mobile consumers are the same. Without understanding consumers in the context of their mobile type as well as their generation, gender, race and ethnicity, motivations, and attitudes toward the press, insight will be limited and so will opportunities to develop effective and innovative strategies to increase the number of news consumers who value the benefits of being informed and are willing to pay for those benefits.

Industry would also benefit from the insight of AEJMC members who are journalism and communication educators, professionals, and graduate students. Some members of the Association for Education in Journalism and Mass Communication are researching the news audience, as well as streams of revenue, and that research is being presented at the annual AEJMC conference and published in scholarly journals such as *Journalism & Mass Communication Quarterly*, *Newspaper Research Journal*, *Journal of Broadcasting & Electronic Media*, and *Mobile Media & Communication*, a recent addition to the academic journal landscape that in 2015 devoted an issue to mobile news engagement around the world (Chan-Olmsted, et al., 2012; Weiss, 2013; Chyi, et al., 2015; Incollingo, 2015; Thorson et al., 2015; Westlund, 2015).

There's also a lot to be learned about the news audience by following #newsengagementday on the first Tuesday of every October, where news consumers enthusiastically share their experiences and insight about engaging with news.

Finally, preparing for what's next also means being ready when mobile-first consumers take over as the dominant mobile consumer in the news media landscape. Based on the *Pediatrics* journal study and observing two-and-a-half-year-old Harper intuitively navigate a smartphone, the dominance of mobile-first consumers will likely happen sooner than we think.

APPENDIX A

METHODOLOGY FOR NEWS FOR A MOBILE-FIRST CONSUMER SURVEY

The News for a Mobile-First Consumer Survey, which was approved by the University of Texas at Austin's Institutional Review Board (IRB), was conducted online during the summer of 2014. I commissioned the Office of Survey Research (OSR) at the University of Texas at Austin to oversee the fieldwork. I wrote the majority of questions on the questionnaire specifically for *News for a Mobile-First Consumer*. I also included questions for the first National News Engagement Day to establish a baseline for news engagement. Additionally, the questionnaire included questions that then-doctoral student Logan Molyneux, now a journalism professor at Temple University, wrote for his Ph.D. dissertation.

The survey questionnaire, which took 10 to 15 minutes to answer, was completed by 1,506 respondents. The sample of online respondents was acquired through Survey Sampling International, an internationally respected survey sampling firm. The source of the online sample was an actively managed panel. Since the panel sample was non-probability rather than random, it was requested that the sample match U.S. demographic statistics. Matching demographics of panel respondents is one method used to strengthen online panels according to AAPOR, the American Association for Public Opinion Research (Online panels, n.d.).

As the Appendix A1 table shows, the sample matched well with U.S. Census data on race and ethnicity and gender. However, the sample education level was higher and the income was lower. Some education differences may be attributed to the Census reporting education for those 25 and older while the *News for a Mobile-First Consumer* statistics are based on all participants who were 18 and older.

The close match with U.S. Census data on race and ethnicity can be attributed to (1) the inclusion of an additional separate question to identify Hispanics or Latinos and (2) the monitoring of the race and ethnicity of panel participants over the course of the survey fieldwork to ensure the racial and ethnic group representation was in line with census data.

First, survey respondents were asked: What is your race or ethnic group? Response choices included Caucasian or White, African American or Black, Hispanic or Latino, Asian American, Native American or Other. The follow-up question to identify Hispanics or Latinos regardless of race was similar to a question used by the *Washington Post* for a survey of digital subscribers: Are you of Hispanic or Latino origin, such as Latin American, Mexican, Puerto Rican or Cuban?

To further validate survey results, the sample results were also compared on wireless-only phones with data from the National Health Interview Survey (NHIS) CDC (Blumberg & Luke, 2013). Of the News for a Mobile-First Consumer Survey participants, 44% lived in households with smartphones and cellphones only. The NHIS data, which was collected in 2013, found that 39.4% lived in homes that only had mobile phones.

To determine generation, respondents were asked what year they were born. Because survey participants were required to be 18 years or older, Millennials younger than 18 were not included. The following is the generational breakdown among survey participants: 29% Millennials (born 1983–1999, with 20% Wave I and 9% Wave II); 30% Generation X (born 1965–1982); 34% Baby Boomers (born 1946–1964); 7% Pre-Baby Boomers (includes Silent Generation born 1925–1945 and Greatest Generation born 1901–1924).

Frequencies, cross-tabulations, factor analysis, and index construction were used to analyze the data and appropriate statistical tests were applied. The vast majority of correlations that are reported in the tables and text are statistically significant; occasionally non-significant correlations are reported because of the insight that can accompany non-significant results. In those cases, the results are labeled accordingly.

Table Appendix A1: Comparison of Mobile-First Consumer Survey Panel Sample and U.S. Census.

	News for Mobile-First Survey Sample	**U.S. Census** (Quick facts, n.d.)
Race and Ethnicity:		
White (Non-Hispanic)	62%	62%
Hispanic/Latino	19%	17%
African American/Black	13%	13%
Asian American	5%	5%
Native American	1%	1%
Gender:		
Male	48%	49%
Female	52%	51%
Education:		
High school degree or more	97%	86%*
College degree or higher	43%	29%*
Median Income	$40–$49,000	$53,046

*Age 25 or older

APPENDIX B

NEWS FOR A MOBILE-FIRST CONSUMER SURVEY QUESTIONNAIRE

1. Approximately how many days in an average week do you get news?
   1. 0 days ______________________________ →(Skip to Q 18.)
   2. 1 day
   3. 2 days
   4. 3 days
   5. 4 days
   6. 5 days
   7. 6 days
   8. 7 days

2. Some people purposely seek news; others get news while doing something else. How many days in an average week do you purposely seek news?
   1. 0 days ______________________________ →(Skip to Q 18.)
   2. 1 day
   3. 2 days
   4. 3 days
   5. 4 days
   6. 5 days
   7. 6 days
   8. 7 days

3. If you get news at least one day per week, from what platform do you usually access it?
   1. Smartphone
   2. Tablet computer
   3. Laptop
   4. E-Reader
   5. TV
   6. Print Newspaper
   7. Print Magazine
   8. Radio
   9. Other ______________________________

The next questions want to know more about your use of print, online, and broadcast news. First, you'll be asked about days per week followed by number of times per day that you get news in print, online, on TV and radio, and a smartphone or tablet.

4. How many *days per week*, on average, do you get news in print?
   1. 0 days ______________________________ →(Skip to Q 6.)
   2. 1 day
   3. 2 days
   4. 3 days
   5. 4 days
   6. 5 days
   7. 6 days
   8. 7 days

5. How many *times per day*, on average, do you get news in print? ________

   ____

6. How many *days per week*, on average, do you get news online?
   1. 0 days ______________________________ →(Skip to Q 8.)
   2. 1 day
   3. 2 days
   4. 3 days
   5. 4 days
   6. 5 days
   7. 6 days
   8. 7 days

7. How many times per day, on average, do you get news online? ________
   ____

8. How many days per week, on average, do you get news on television?
   1. 0 days ______________________ →(Skip to Q 10.)
   2. 1 day
   3. 2 days
   4. 3 days
   5. 4 days
   6. 5 days
   7. 6 days
   8. 7 days

9. How many times per day, on average, do you get news on television?
   ____

10. How many days per week, on average, do you get news on radio?
    1. 0 days ______________________ →(Skip to Q 12.)
    2. 1 day
    3. 2 days
    4. 3 days
    5. 4 days
    6. 5 days
    7. 6 days
    8. 7 days

11. How many times per day, on average, do you get news on radio?
    ____

12. How many days per week, on average, do you get news on a smartphone or tablet?
    1. 0 days ______________________ →(Skip to Q 14.)
    2. 1 day
    3. 2 days
    4. 3 days
    5. 4 days
    6. 5 days
    7. 6 days
    8. 7 days

13. How many times per day, on average, do you get news on your smartphone or tablet? ______

14. Think of the last time you used each of the following media for news. About how long did you spend getting news that time?

	You don't use this medium	10 minutes or less	About 11–20 minutes	About 21–30 minutes	About 31–40 minutes	About 41–50 minutes	About 51–60 minutes	More than an hour
A. In print	0	1	2	3	4	5	6	7
B. Online on a laptop or desktop computer	0	1	2	3	4	5	6	7
C. On television	0	1	2	3	4	5	6	7
D. By radio	0	1	2	3	4	5	6	7
E. On a smartphone or tablet	0	1	2	3	4	5	6	7

These questions want to know how often you pay attention to various categories of news.

15. Do you often, sometimes, occasionally or never pay attention to the following categories of news:

	Often	Sometimes	Occasionally	Never
A. International News	1	2	3	4
B. National News	1	2	3	4
C. Local News	1	2	3	4
D. Celebrity & Entertainment News	1	2	3	4
E. Sports News	1	2	3	4
F. Weather News	1	2	3	4
G. Health News	1	2	3	4
H. Business News	1	2	3	4
I. Arts and Culture News	1	2	3	4
J. Politics	1	2	3	4
K. Government News	1	2	3	4
L. Natural Disaster News	1	2	3	4
M. Technology News	1	2	3	4
N. Election News	1	2	3	4
O. Accidents	1	2	3	4
P. Crime News	1	2	3	4
Q. Consumer News & Personal Finance	1	2	3	4
R. Environment News	1	2	3	4
S. Breaking News	1	2	3	4
T. Opinions about the News	1	2	3	4

16. In general, what grade, from A to F, would you give the news media for its news coverage?
    A. A
    B. B
    C. C
    D. D
    E. F

17. What is the main reason for your grade?

____

The next questions will ask about social media.

18. Are you currently on Facebook?
    1. Yes
    2. No __________ →(Skip to Q 20.)

19. From 0% to 100%, approximately what percent of the time you spend on Facebook on a typical day is devoted to engaging with news? __________

    ____

20. Are you currently on Instagram?
    1. Yes
    2. No

21. Are you currently on Twitter?
    1. Yes
    2. No ____ → (Skip to Q 31.)

22. Approximately how many days in an average week are you using Twitter?
    1. 0 days __________________________ →(Skip to Q 31.)
    2. 1 day
    3. 2 days
    4. 3 days
    5. 4 days
    6. 5 days
    7. 6 days
    8. 7 days

23. What do you primarily do on Twitter?

    ____

24. From 0% to 100%, approximately what percent of the time you spend on Twitter on a typical day is devoted to engaging with news?

    ____

25. How do you usually access Twitter?
    1. Mobile app
    2. Browser on a smartphone or tablet
    3. Browser on a laptop
    4. Other ______________________________

26. Do you regularly follow a journalist or news organization on Twitter?
    1. Yes
    2. No

27. What journalist or news organization do you regularly follow on Twitter?

    ____

28. How often do you click on a news story that you see tweeted?
    1. Often
    2. Sometimes
    3. Occasionally
    4. Never

29. How often do you retweet a news story?
    1. Often
    2. Sometimes
    3. Occasionally
    4. Never

30. How often do you click on a news story that is trending?
    1. Often
    2. Sometimes
    3. Occasionally
    4. Never

These questions are about mobile devices, including smartphones and tablet computers.

31. Which, if any, mobile devices do you own?

	Yes	No
1. Smartphone such as an iPhone or Samsung Galaxy	1	2
2. Tablet computer such as an iPad	1	2
3. Cellphone (excluding iPhone, Samsung or other Android smartphone)	1	2
4. E-reader	1	2

(Skip to Q 42 if no mobile devices owned.)

32. What do you most use your mobile device for? ______________________

____

33. Which of the following social media apps do you have on your mobile device?

	Yes	No
A. Facebook	1	2
B. Twitter	1	2
C. YouTube	1	2
D. LinkedIn	1	2
E. Instagram	1	2
F. Other	1	2

34. Approximately how many apps on your mobile device are for news?

____

35. Do you use your mobile device to engage with news?
   1. Yes
   2. No (Skip to Q 42.)

36. How do you usually engage with news on your mobile device?
   1. App
   2. Browser
   3. Twitter
   4. Facebook
   5. Notification

6. Breaking news alert
7. Search engine
8. Other ____________________________

37. From what site are you primarily getting the news you engage with? ____________________

38. What news app, if any, do you use most often on your mobile device? ________________

39. From 0% to 100%, approximately what percent of the time that you spend on your mobile device on a typical day is devoted to engaging with news? ________________________________

40. What grade, from A to F, would you give the presentation of news on your mobile device?
    1. A
    2. B
    3. C
    4. D
    5. F

41. What is the main reason for your grade? ______________________ ____

42. Here are some things that have been said about the news and journalism. Please indicate your level of agreement or disagreement with the statement with 1 being strongly disagree and 5 being strongly agree.

	Strongly Disagree				Strongly Agree
A. You enjoy keeping up with the news.	1	2	3	4	5
B. We all have a duty to keep ourselves informed about news and current events.	1	2	3	4	5
C. Keeping up with the news is a waste of time.	1	2	3	4	5
D. It's important for the news media to be a watchdog of powerful people and the government.	1	2	3	4	5

E. The news does a poor job of providing the context to understand a story.	1	2	3	4	5
F. The press in America has too much freedom to do what it wants.	1	2	3	4	5
G. You were raised to believe being informed of the news is important.	1	2	3	4	5
H. Most news is biased.	1	2	3	4	5
I. Being informed of the news makes you feel empowered.	1	2	3	4	5
J. The news media should be more helpful and caring like a good neighbor.	1	2	3	4	5
K. All news on the Internet should be free to the public.	1	2	3	4	5
L. The news media care little about people like you.	1	2	3	4	5
M. Most of the people who are important to you value keeping up with the news.	1	2	3	4	5
N. You depend on the news to help you with your daily life.	1	2	3	4	5
O. News organizations are more concerned about being first than being right.	1	2	3	4	5
P. News organizations are willing to admit their mistakes.	1	2	3	4	5
Q. News stories are often inaccurate.	1	2	3	4	5
R. Mobile devices make it easy for you to keep up with the news.	1	2	3	4	5
S. News organizations protect democracy more than hurt democracy.	1	2	3	4	5
T. You depend on social media to help you with your daily life.	1	2	3	4	5
U. The news projects an accurate representation of the constituent groups in U.S. society.	1	2	3	4	5
V. Reading news on a smartphone is frustrating.	1	2	3	4	5

43. How much do you trust the following as sources for news?

	A lot	Some	None
A. Local TV news	1	2	3
B. Cable News	1	2	3
C. Twitter	1	2	3
D. Network TV news	1	2	3
E. Local newspaper, including online and print	1	2	3
F. Facebook	1	2	3
G. Friends & Friends	1	2	3
H. Journalists	1	2	3
I. National Newspapers	1	2	3
J. Google	1	2	3
K. Bloggers	1	2	3
L. Government websites	1	2	3

44. What is the most important problem facing this country today? ________

____

45. What grade, from A to F, would you give the news media for its coverage of the most important problem facing this country today?
   1. A
   2. B
   3. C
   4. D
   5. F

These questions are about news subscriptions and how much, if anything, you spend on those subscriptions.

46. Do you pay for any news subscriptions?
   1. Yes
   2. No (Skip to Q 49.)

47. Which of the following news subscriptions do you have?

	Yes	No
A. Printed news products (i.e., newspaper, magazine)	1	2
B. Digital (news website, online, mobile)	1	2

48. About how much do you pay per month for your news subscription(s)?

____

49. How often do you buy single copies of newspapers or news magazines from a newsstand?
    1. Often
    2. Sometimes
    3. Occasionally
    4. Never

The next few questions are about your civic awareness and participation.

50. What freedoms are listed in the First Amendment to the U.S. Constitution?

____

51. Did you vote in the 2012 U.S. Presidential Election?
    1. Yes
    2. No

52. Do you intend to vote in the 2014 mid-term election?
    1. Definitely will vote
    2. Probably will vote
    3. Probably will not vote
    4. Definitely will not vote

53. Do you intend to vote in the 2016 presidential election?
    1. Definitely will vote
    2. Probably will vote
    3. Probably will not vote
    4. Definitely will not vote

54. In the past year, have you contacted an elected official in any way (via mail, phone, email, social media, etc.)?
    1. Yes
    2. No

55. In the past year, have you attended a public meeting or hearing to discuss community problems?
    1. Yes
    2. No

56. In the past year, have you signed or shared a petition (in person or online)?
    1. Yes
    2. No

57. In the past year, have you donated money to a campaign or political cause?
    1. Yes
    2. No

58. In the past year, have you donated money to or volunteered for a charity or community organization?
    1. Yes
    2. No

59. Generally speaking, do you identify yourself as a:
    1. Democrat
    2. Republican
    3. Independent
    4. Other ____________________

60. Do you consider yourself:
    1. Conservative
    2. Middle of the road
    3. Liberal
    4. Other ______________

61. Do you consider yourself a member of the Tea Party?
    1. Yes
    2. No

62. What year were you born? __________________________

63. Are you?
    1. Male
    2. Female

64. What is the highest level of education that you've completed?
    1. Some High School or Less
    2. High School Degree
    3. Some College or Technical School Degree

4. College Graduate
5. Some Graduate or Professional School
6. Masters, M.D. or Doctorate
7. Other ______________

65. What is your race or ethnic group?
    1. Caucasian or White
    2. African American or Black
    3. Hispanic or Latino
    4. Asian American
    5. Native American
    6. Other _________________

66. Are you of Hispanic or Latino origin, such as Latin American, Mexican, Puerto Rican, or Cuban?
    1. Yes
    2. No

67. Approximately, what is your household income?
    1. Under $20,000
    2. $20–$29,000
    3. $30–$39,000
    4. $40–$49,000
    5. $50–$59,000
    6. $60–$69,000
    7. $70–$79,000
    8. $80–$89,000
    9. $90–$99,000
    10. $100,000 or more

68. Some people have both a landline and a smartphone. Others have one or the other. What about you? Do you own:
    1. Landline phone only
    2. Cellphone or smartphone only
    3. Landline and cellphone
    4. Other ____________________

69. What is your zip code? ____________________________________

Finally, I have a few more questions about some topics that have been in the news.

70. Which federal agency's documents did Edward Snowden leak to the news media?
    1. FBI
    2. CIA
    3. NSA
    4. The Federal Reserve

71. What does the term "Common Core" refer to?
    1. School curriculum standards for language and math
    2. The military's code of conduct
    3. A newly developed microprocessor
    4. Abdominal exercises

72. Did the Affordable Care Act meet, exceed, or fall short of its target number of enrollees?
    1. Met the target number
    2. Exceeded the target number
    3. Fell short of the target number

73. The U.S. unemployment rate increased during the 2008 recession. How does today's unemployment rate compare?
    1. It is lower than during the recession.
    2. It is about the same as during the recession.
    3. It is higher than during the recession.

74. Of what country is Vladimir Putin president?
    1. Turkey
    2. Ukraine
    3. Russia
    4. Syria

Thank you for answering the questions.

APPENDIX C

METHODOLOGY FOR NEWS FOR A MOBILE-FIRST CONSUMER SUPPLEMENTAL SURVEY

The News for a Mobile-First Supplemental Survey was conducted in the summer of 2015. It, too, was an online survey but the source of the sample was Amazon's mTurk, which provided a pool of prospective respondents who could be screened to qualify to answer the questionnaire. Because the goal of the supplemental survey was to provide baseline social media comparison insight about Snapchat as a platform for news from mobile consumers' perspective, there was no effort to get participants to represent the U.S. Census on race and ethnicity and it didn't.

The supplemental survey questionnaire was completed by 1,166 respondents. Results showed a 50–50 gender split and that 76% of respondents were white, 8% were African American or black, 7% were Hispanic or Latino, 8% were Asian American, and less than 1% were Native American. Additionally, 55% of supplemental survey participants had at least a college degree, 55% reported a household income of less than $50,000, and 68% lived in households with a smartphone or cellphone only.

Based on year born, the following is the generational breakdown: 54% Millennials (born 1983–1999), 33% Generation X (born 1965–1982), 13% Baby Boomers (born 1946–1964), 1% Pre-Baby Boomers (Includes Silent Generation born 1925–1945 and Greatest Generation born 1901–1924).

In addition to questions from the original News for a Mobile-First Consumer Survey, the supplemental survey questionnaire included new questions on social media with special attention paid to Snapchat. Additionally, there were new questions on mobile news consumption and civic engagement, which then-doctoral student Logan Molyneux, who conducted the survey, used in his dissertation, which I chaired (Molyneux, 2015).

APPENDIX D

NEWS FOR A MOBILE-FIRST SUPPLEMENTAL SURVEY QUESTIONNAIRE

1. How many days in an average week do you get news?
    1. 1 day
    2. 2 days
    3. 3 days
    4. 4 days
    5. 5 days
    6. 6 days
    7. 7 days
    8. I don't usually get any news. (Skip to Q37.)

2. Today people get news from a variety of devices and platforms. Which of the following devices or platforms do you use the most to get news? (Check all that apply)
    1. Laptop or desktop computer
    2. Smartphone such as iPhone or Android
    3. Tablet computer such as an iPad
    4. Television
    5. Radio
    6. Printed newspaper or magazine
    7. Other_____________

2a. (Using checked items from Q2) On which device or platform do you get the most news?

____

3. Some people briefly check in on news; others get news in longer sittings. In general, which best describes how you usually get news?
   1. You briefly check in on news
   2. You get news in longer sittings
   3. You get news in both longer sittings and brief check-ins
   4. Other

4. When you do get news, about how many minutes do you usually spend on news before moving on to something else? ______________________

5. Below are some statements describing ways that people get news. How much do these descriptions apply to how you get news?

	Not at all	A little	Some	A lot
A. You get news in bits and pieces throughout the day.	1	2	3	4
B. You get news in longer sittings of 20 minutes or more.	1	2	3	4
C. When you get news, you focus on it for periods of 20 minutes or more.	1	2	3	4
D. You generally get news in short bursts.	1	2	3	4
E. You usually read, watch or listen to only one or two news stories before moving on to something else.	1	2	3	4
F. You usually spend more than 20 minutes getting news before moving on to something else.	1	2	3	4

These next questions ask more about your use of different platforms for news, including print, television, computer, radio, and mobile devices. You'll be asked about how many days per week you get news on that platform, and then your use of that platform yesterday.

6. About how many days per week, on average, do you read a printed newspaper?
   1. 1 day

2. 2 days
3. 3 days
4. 4 days
5. 5 days
6. 6 days
7. 7 days
8. I don't usually get news in print (Skip to Q10.)

7. Did you get a chance to read a printed newspaper yesterday?
   1. Yes
   2. No (Skip to Q10.)

8. About how much time did you spend reading the newspaper yesterday?
   Hours: _____ Minutes: ____

9. Thinking of the time you spent reading the newspaper yesterday, was it all at once, or spread out over the course of the day?
   1. All at once
   2. Spread out over the course of the day (Show Q9a.)

9a. Thinking of the last time you read the newspaper yesterday, about how long did you spend getting news before moving on to something else?
   Hours: _____ Minutes: ____

10. About how many days per week, on average, do you watch the news on TV?
    1. 1 day
    2. 2 days
    3. 3 days
    4. 4 days
    5. 5 days
    6. 6 days
    7. 7 days
    8. I don't usually watch the news on TV (Skip to Q14.)

11. Did you get a chance to watch the news on TV yesterday?
    1. Yes
    2. No (Skip to Q14.)

12. About how much time did you spend watching the news on TV yesterday? Hours: _____ Minutes: ____

13. Thinking of the time you spent watching the news on TV yesterday, was it all at once, or spread out over the course of the day?
    1. All at once
    2. Spread out over the course of the day (Show Q13a.)

13a. Thinking of the last time you got news on TV yesterday, about how long did you spend getting news before moving on to something else? Hours: _____ Minutes: ____

14. About how many days per week, on average, do you get news online using a laptop or desktop computer?
    1. 1 day
    2. 2 days
    3. 3 days
    4. 4 days
    5. 5 days
    6. 6 days
    7. 7 days
    8. I don't usually get news on a computer (Skip to Q18.)

15. Did you get a chance to get news online using a laptop or desktop computer yesterday?
    1. Yes
    2. No (Skip to Q18.)

16. About how much time did you spend getting news on a laptop or desktop computer yesterday? Hours: _____ Minutes: ____

17. Thinking of the time you spent getting news on a laptop or desktop computer yesterday, was it all at once, or spread out over the course of the day?
    1. All at once
    2. Spread out over the course of the day

17a. Thinking of the last time you got news on a laptop or desktop computer yesterday, about how long did you spend getting news before moving on to something else? Hours: _____ Minutes: ____

18. About how many days per week, on average, do you listen to news on the radio?
    1. 1 day
    2. 2 days
    3. 3 days
    4. 4 days
    5. 5 days
    6. 6 days
    7. 7 days
    8. I don't usually listen to news on the radio (Skip to Q22.)

19. Did you get a chance to listen to news on the radio yesterday?
    1. Yes
    2. No (Skip to Q22.)

20. About how much time did you spend listening to news on the radio yesterday? Hours: _____ Minutes: ____

21. Thinking of the time you spent listening to news on the radio yesterday, was it all at once, or spread out over the course of the day?
    1. All at once
    2. Spread out over the course of the day

21a. Thinking of the last time you got news on the radio yesterday, about how long did you spend getting news before moving on to something else? Hours: _____ Minutes: ____

22. About how many days per week, on average, do you get news on a smartphone, such as an iPhone or Android?
    1. 1 day
    2. 2 days
    3. 3 days
    4. 4 days
    5. 5 days
    6. 6 days
    7. 7 days
    8. I don't usually get news on a smartphone (Skip to Q26.)

23. Did you get a chance get news on a smartphone (such as an iPhone or Android) yesterday?
    1. Yes
    2. No (Skip to Q26.)

24. About how much time did you spend getting news on a smartphone yesterday? Hours: _____ Minutes: ____

25. Thinking of the time you spent getting news on a smartphone yesterday, was it all at once, or spread out over the course of the day?
    1. All at once
    2. Spread out over the course of the day

25a. Thinking of the last time you got news on a smartphone yesterday, about how long did you spend getting news before moving on to something else? Hours: _____ Minutes: ____

26. About how many days per week, on average, do you get news on a tablet, such as an iPad or Kindle?
    1. 1 day
    2. 2 days
    3. 3 days
    4. 4 days
    5. 5 days
    6. 6 days
    7. 7 days
    8. I don't usually get news on a tablet (Skip to Q30.)

27. Did you get a chance get news on a tablet (such as an iPad or Kindle) yesterday?
    1. Yes
    2. No (Skip to Q30)

28. About how much time did you spend getting news on a tablet yesterday? Hours: _____ Minutes: ____

29. Thinking of the time you spent getting news on a tablet yesterday, was it all at once, or spread out over the course of the day?
    1. All at once
    2. Spread out over the course of the day (Show Q29a.)

29a. Thinking of the last time you got news on a tablet yesterday, about how long did you spend getting news before moving on to something else? Hours: _____ Minutes: ____

30. How do you usually get news on your mobile device?
    1. App
    2. Browser
    3. Twitter
    4. Facebook
    5. Notification
    6. Breaking news alert
    7. Search engine
    8. Snapchat
    9. Other ___________

These questions want to know how often you pay attention to various categories of news.

31. Do you often, sometimes, occasionally, or never pay attention to the following categories of news:

	Never	Occasionally	Sometimes	Often
A. International News	1	2	3	4
B. National News	1	2	3	4
C. Local News	1	2	3	4
D. Celebrity & Entertainment News	1	2	3	4
E. Sports News	1	2	3	4
F. Weather News	1	2	3	4
G. Health News	1	2	3	4
H. Business News	1	2	3	4
I. Political News	1	2	3	4
J. Crime, Accident and Disaster News	1	2	3	4
K. Technology News	1	2	3	4
L. Breaking News	1	2	3	4
M. Environment News	1	2	3	4

These questions are about mobile devices, including smartphones and tablet computers.

32. Which, if any, devices do you own?

	Yes	No
A. Smartphone such as an iPhone or Samsung Galaxy	1	2
B. Tablet computer such as an iPad	1	2
C. Cellphone (excluding iPhone, Samsung or other Android smartphone)	1	2
D. E-reader	1	2

33. Which of the following social media apps do you have?

	Yes	No
A. Facebook	1	2
B. Twitter	1	2
C. Google+	1	2
D. Instagram	1	2
E. YouTube	1	2
F. Snapchat	1 (Answer Q34.)	2 (Skip to Q35.)
G. Other	1	2

34. What do you primarily do on Snapchat?

____

35. How often do you see news on the following mobile apps? (If you do not have the app, please click: 5, Don't have app.)

	Often	Sometimes	Occasionally	Never	(Don't have app.)
A. Facebook	1	2	3	4	5
B. Twitter	1	2	3	4	5
C. Google+	1	2	3	4	5
D. Instagram	1	2	3	4	5
E. YouTube	1	2	3	4	5
F. Snapchat	1	2	3	4	5

(If F is 4 or 5, skip to Q38)

36. How much attention have you paid to the news you've seen on Snapchat?
   1. A lot
   2. Some
   3. None

37. What is your opinion of the news you've seen on Snapchat?

____

The next few questions are about your civic awareness and participation.

38. Did you vote in the 2012 U.S. Presidential Election?
    1. Yes
    2. No

39. In the past year, have you contacted an elected official in any way (via mail, phone, email, social media, etc.)?
    1. Yes
    2. No

40. In the past year, have you attended a public meeting or hearing to discuss community problems?
    1. Yes
    2. No

41. In the past year, have you signed or shared a petition (in person or online)?
    1. Yes
    2. No

42. In the past year, have you donated money to a campaign or political cause?
    1. Yes
    2. No

43. In the past year, have you donated money to or volunteered for a charity or community organization?
    1. Yes
    2. No

[News Knowledge; randomize response order]

44. Which major U.S. city recently voted to raise the minimum wage to $15?
    1. Los Angeles
    2. Chicago
    3. New York
    4. Seattle

45. The largest auto recall in history was recently announced because of defective:
    1. Airbags
    2. Ignition switches
    3. Exhaust valves
    4. Seatbelts

46. To comply with the health care law, most Americans must now indicate they have health insurance coverage when they:
    1. File their taxes
    2. Receive a driver's license
    3. Register to vote
    4. Apply for a job

47. The United States recently re-established diplomatic relations with which country?
    1. Cuba
    2. Russia
    3. North Korea
    4. Yemen

48. The U.S. Court of Appeals recently ruled that it is illegal for which federal agency to collect Americans' phone records in bulk?
    1. NSA
    2. FBI
    3. CIA
    4. TSA

49. Do you generally think of yourself as a Republican, a Democrat, or an independent?

Strong Democrat				Independent				Strong Republican		
1	2	3	4	5	6	7	8	9	10	11

50. Think about the people you have talked to regarding politics or public affairs. During the past week, about how many total people have you talked to face-to-face or over the phone about politics or public affairs?

____

51. Still thinking about the people that you have talked to about politics or public affairs during the past week, about how many people would you say you have talked to via the Internet, including texting, e-mail, chat rooms, social networking sites and micro-blogging sites? ________________

52. How much do you agree or disagree with the following statements about public life?

	A lot	Some	A little	Not at all
A. People like me can influence government.	1	2	3	4
B. I consider myself qualified to participate in politics.	1	2	3	4
C. I have a good understanding of the important political issues facing our country.	1	2	3	4
D. People like me don't have any say in what the government does.	1	2	3	4
E. No matter whom I vote for, it won't make a difference.	1	2	3	4
F. I don't think public officials care much about what people like me think	1	2	3	4

53. What year were you born? ________________________

54. Are you?
   1. Male
   2. Female

55. What is the highest level of education that you've completed?
   1. Some High School or Less
   2. High School Degree
   3. Some College or Technical School Degree
   4. College Graduate
   5. Some Graduate or Professional School
   6. Masters, M.D. or Doctorate
   7. Other _______________

56. What is your race or ethnic group?
    1. Caucasian or White
    2. African American or Black
    3. Hispanic or Latino
    4. Asian American
    5. Native American
    6. Other ________________

57. Are you of Hispanic or Latino origin, such as Latin American, Mexican, Puerto Rican or Cuban?
    1. Yes
    2. No

58. Approximately, what is your household income?
    1. Under $20,000
    2. $20–$29,000
    3. $30–$39,000
    4. $40–$49,000
    5. $50–$59,000
    6. $60–$69,000
    7. $70–$79,000
    8. $80–$89,000
    9. $90–$99,000
    10. $100,000 or more

59. Some people have both a landline and a smartphone. Others have one or the other. What about you? Do you own:
    1. Landline phone only
    2. Cellphone or smartphone only
    3. Landline and cellphone
    4. Other ________________

60. What is your zip code? ________________________________

Those are all of my questions. Thank you for your participation.

APPENDIX E

ETHICAL GUIDELINES FOR JOURNALISTS USING SOCIAL MEDIA

Social media are not just integrated into the news media landscape; they also are essential platforms for journalists who use them to gather, report, and distribute news. That's why ethical guidelines for social media are included in ONA's Build Your Own Ethics Code project, which as the project leader Associated Press Deputy Managing Editor/Standards Editor Tom Kent said "grew out of an open workshop at the Online News Association annual conference in Atlanta in October 2013, where journalists expressed a need for codes that fit the world we live in and report on, and the tools we use" (Kent, n.d.).

Consumers' high use of social media but low trust level of social media as platforms for news, as learned in *News for a Mobile-First Consumer*, underscores that journalists and news organizations must be guided by the highest ethical standards when engaging with social media. This social network module, written by AP's Eric Carvin for ONA's Build Your Own Ethics Code project provides journalists, journalism educators, students of journalism, and news consumers with ethical guidelines and insight about how journalism should be practiced when social media are part of the process (Build your ethics, n.d.).

Social Networks

This module for ONA's Build Your Own Ethics Code project (Carvin, n.d., reprinted with permission) addresses these ethical questions:

- Can journalists use social networks to express opinions or advocate for causes?
- What about retweeting, reposting or reblogging the opinions of others?
- Is that an acceptable practice for a journalist, and if so, how should it be handled?
- Is it wise for a journalist to get in angry exchanges with the public on social networks?
- To what degree should journalists be required to identify themselves when using social networks?
- What level of verification is needed for a journalist to share news over social networks?
- Are there safety or security issues to consider when posting on social networks?
- Is it OK for a journalist to friend or follow a source?
- Should social network postings be considered private, necessitating permission before quoting them?
- What's an ethical way to correct errors made on social networks?

Social networks raise a variety of issues (http://mediamorals.org/twitter-and-journalism-three-continuing-ethical-pressure-points/) for news organizations. Note that this section focuses strictly on the use of social networks for engagement and distribution of information and content. See the separate section on "User-generated content" for issues related to social newsgathering.

Advocacy and Expressions of Opinion by Journalists

Social networking profiles can give journalists an opportunity to show what they're like as people—people with opinions, feelings, and individual experiences. At the same time, a journalist who tweets or posts is a representative of his or her news organization. What does this mean for expressions of opinion?

Impartial News Organizations

If you're at a traditional news organization that focuses on reporting the news in an unbiased way, your journalists should, at a minimum, avoid expressing opinions on contentious issues. Opinionated tweets and posts can easily become associated with your news organization as a whole and can damage its reputation as an impartial outlet. That, in turn, can harm relationships with customers and sources.

Some details you'll need to decide on: Does this practice also apply to your non-news staff? They could also be associated with your organization if they do something controversial. Can a journalist express opinions on a topic she doesn't cover? It's usually best to avoid the most controversial topics no matter what, though there might otherwise be some wiggle room. But think about whether a journalist may, some day, cover different topics, or if one staffer's opinions might affect another staffer's ability to report the news.

What about opinions on sports and entertainment? Opinions in these areas might be acceptable for journalists who don't cover them, though gushing praise or angry vitriol may still be an issue. And think about whether you need exceptions for certain journalists, such as columnists or bloggers who express opinions as part of their jobs.

Point-of-View News Organizations

What if your news organization openly professes certain points of view? In that case, opinion by your journalists may not be a problem—in fact, it may fit in with the character of your operation. You may need to consider this on an issue-by-issue basis, depending on the focus of your news organization.

Even if your organization encourages staff members to share opinions, consider whether you want to place any limitations, such as barring personal attacks or foul language, or discouraging running arguments on social media.

What about Retweeting, Reblogging, and Reposting?

In situations where a journalist is asked to refrain from expressing opinions, can she retweet, reblog, reshare, or otherwise repost the opinions of others? The best practice is to append some material on such posts to put them into context, so the re-posts aren't seen as an endorsement of the original opinion.

Some news organizations find it acceptable to do a straight retweet or resharing of something opinionated, as long as one's social profile notes that retweets and other similar posts don't constitute endorsements; others note that these profile details are rarely read by others, and that such a disclaimer becomes meaningless when a journalist repeatedly re-shares the same sorts of opinions.

Angry Exchanges

In general, journalists should avoid getting into angry exchanges with fellow journalists, sources, members of the public or readers/viewers on any topic. We have a responsibility to treat people with respect. There may be some exceptions, though—journalists who are paid to be opinionated might benefit their news organizations by mixing it up a bit, though they should be careful not to cross over into abuse. And journalists at point-of-view news organizations can certainly engage in civil debate on topics of interest, which can encourage discussion and serve to inform viewers and readers.

Identifying Yourself

In general, journalists should clearly identify themselves by their name and organization in the biographical material of any social network they use. You'll want to give some thought to whether there should be any exceptions to this—in particular, whether a journalist might be allowed to create an account to go undercover in pursuit of an investigative story. This can be a dangerous practice, given the importance we journalists place in getting accurate identities from our sources. So if this happens at all, it should be done rarely, in consultation with the organization's top news executives, and only in pursuit of high-impact news that can't be reported another way.

Using Social Networks to Share News Developments

How certain do you do need to be about a news development to share it on a social network, whether from an institutional account or one belonging to an individual journalist? An ethical approach to social journalism demands some standards, though the exact approach can reasonably vary. Here are some considerations for various types of shareable information.

Rumors/hearsay: Many journalists believe they should report news only from legitimate sources, avoiding rumors—whether they're found on a social network, overheard in a public place, or discovered in some other way. An exception might be made if the existence of such rumors is itself a newsworthy phenomenon (e.g., if unfounded rumors prompt an elected official to hold a news conference to set the record straight). An exception could also be justified for journalists who specialize in using social networks to verify rumors through crowdsourcing. In that case, material that is only rumor should be clearly identified as such. This is explored further by the Ethics Advice Line for Journalists.

Reports by others: You'll need to make a decision about whether (and under what circumstances) your news organization or journalists might tweet or post information reported by another news organization or journalist. Some news operations already have standards for when they might pick up information from another news organization; if so, this might also inform your policy on what you might retweet or otherwise reshare. Maybe you'd only do this if you're familiar with the originating reporter. Or maybe only if the source is identified and is credible. Maybe you wouldn't do it at all, or only if your news organization already picked up the information for a story. Have a plan before news breaks.

Information tweeted/posted by a primary source: What if, for example, a prominent public figure, government agency or business directly shares information over a social network? In most cases, this would be reasonable to reshare, with some caveats:

- You need to be certain you know that the account is authentic. Some news organizations use the presence or absence of a "verified" badge as a guide, but others do their own verification, noting that such badges have been wrongly handed out before. The best approach is to ask an official representative whether the account is real (and to do so before ("before" in italics) big news breaks).
- When you verify an account, find out what you can about how the account is run; you'll be able to make better decisions if you know whether a celebrity or a staff member, for example, is doing the tweeting.
- Keep an eye out for red flags indicating someone's account may have been compromised, like uncharacteristic posts, surprising grammatical errors or sharply worded political opinions.

Observations and color from the scene: This is generally fair game, though see the separate note on safety below.

Independently verified information: This is fine to share on social networks from an ethical perspective—though you'll need to factor in your news organization's business considerations when deciding whether to break news on social or wait until it's distributed to customers some other way. Consider whether the limitations of a social network allow you to report information in context. Use multiple tweets if needed to provide context or answer questions that arise.

Safety and security: Regardless of what sort of news organization you represent, you need to consider any safety issues that might arise from your decisions about what to tweet or post. Sources, members of the public, story subjects, and fellow journalists could be put in danger by a careless tweet or post, such as one that identifies someone's location in a war zone or that identifies a source who put his life at risk by sharing information with you. If someone in the public is posting to social media from the scene of a crime or disaster, encourage them to be careful of their safety. When in doubt, skip the tweet or posts.

Friending and Following

Is it OK for a journalist to friend or otherwise follow a source? Generally, yes; it's a reasonable way to keep up with what they're posting, which is central to your job. However, if you work for an impartial news organization, make sure you follow and at least attempt to friend people on both sides of any contentious issue to avoid perceptions of bias. And think carefully before friending or following a source who's giving you information anonymously. This action might reveal a source's identity. Ask the source if it's OK, and when in doubt, steer clear.

Within your news organization, consider whether it's wise for employees to issue friend requests to one another. Some news organizations may decide that people shouldn't friend colleagues working below them because of the pressure the more junior employee would feel to say yes. There isn't usually an issue in the other direction. At other news organizations, being connected on a social network like Facebook or LinkedIn may be central to the operation, so open friending might be encouraged. Following on a channel like Twitter is generally fine within the newsroom, as a follow is a one-way action that doesn't require approval.

Are Any Posts "Private"?

Many journalists consider that anything posted on a social network is public. But even the contents of password-protected sites and streams might be revealed by someone with access. As a result, some reporters believe people posting on social should have no expectation of privacy, whatever security settings they set up or however few followers or friends they have.

Other reporters have a more conservative approach. If they wish to quote a post made to a very small number of people, they may feel an obligation to contact the poster and at least let him or her know that they plan to quote the post. They also feel a need to make sure that any post they quote reflects the general tone of the person's other postings.

In any case, most journalists agree that when quoting social network postings, we should be clear that the quote comes from a social network, and which one.

Errors and Corrections

Regardless of the nature of a news organization, journalists should do everything they can to avoid errors in their tweets and posts—a mistake made there shouldn't be seen as somehow less serious than one published in another way.

That said, mistakes will be made, and they need to be corrected. The best practice is to put out a new post with the correct information, noting that the other post was corrected and exactly what was wrong. Some social networks, like Facebook, give you the option of revising a post after publication and offer readers an easy way to see the edit history. You'll need to decide how and whether to use this sort of feature. Some organizations leave the original post up, for transparency's sake, if it can't be edited, unless it's determined to be defamatory or otherwise legally problematic. Other organizations favor deleting inaccurate posts that can't be edited, rather than leaving them online to be read or shared without the correcting post. You also might want to consider whether your responsibility to correct an error includes direct communication to people who have retweeted, liked, or shared your erroneous post. One good approach is to edit any post that needs a correction, and to put up an additional post or comment explaining what was changed and why.

Note: The main author of this module is Eric Carvin of the Associated Press.

See also the modules on "Accuracy," "Corrections," and "Removing material from your archives" (Build your ethics, n.d.).

Additional Resources

"How to verify information from tweets: Check it out," *The Buttry Diary*, January 21, 2013.

REFERENCES

A free and responsible press: A general report on mass communication: newspapers, radio, motion pictures, magazines, and books. (1947). Chicago: Hutchins Commission on Freedom of the Press.

A herstory of the #BlackLivesMatter movement. (n.d.). Retrieved from http://blacklivesmatter.com/herstory/

About moments. (n.d.). Retrieved from https://support.twitter.com/articles/20174546?lang=en

About NAA. (n.d.). Retrieved from http://www.naa.org/About-NAA.aspx

AEJMC (Producer). (n.d.). News Engagement Day. Retrieved from newsengagement.org

AEJMC (Producer). (2014a). AEJMC 2014 in Montreal: Paula Poindexter's Presidential Panel Session. Retrieved from https://www.youtube.com/watch?v=BtJ1aitCnPs

AEJMC (Producer). (2014b). The 2014 AEJMC Keynote featured Jill Abramson, the former executive editor of the *New York Times*. Retrieved from https://www.youtube.com/watch?v=BNwCMppP5z8

American Indian and Alaska Native heritage month. (2012). Retrieved from https://www.census.gov/newsroom/releases/archives/facts_for_features_special_editions/cb12-ff22.html

Asian/Pacific American heritage month. (2015). Retrieved from https://www.census.gov/content/dam/Census/newsroom/facts-for-features/2015/cb15-ff07_asian_american-pacific_islanders.pdf

The Associated Press stylebook and briefing on media law. (2013). New York: Basic Books.

Barry, D., Barstow, D., Glater, J. D., Liptak, A., & Steinberg, J. (2003). Correcting the record; Times reporter who resigned leaves long trail of deception. Retrieved from http://www.nytimes.com/2003/05/11/us/correcting-the-record-times-reporter-who-resigned-leaves-long-trail-of-deception.html?pagewanted=all

Batsell, J. (2015). *Engaged journalism: Connecting with digitally empowered news audiences*. New York: Columbia University Press.

Battaglio, S., & Halper, E. (2015). ABC's Stephanopoulos under fire for failing to disclose donations. Retrieved from http://www.latimes.com/entertainment/envelope/cotown/la-et-ct-abc-george-stephanopoulos-clinton-donation-20150515-story.html

Beaujon, A. (2014). *New York Times* will launch new subscription products April 2. Retrieved from http://www.poynter.org/news/mediawire/244803/new-york-times-will-launch-new-subscription-products-april-2/

Berelson, B. (1949). What "missing the newspaper" means. In P. F. Lazarsfeld & F. N. Stanton (Eds.), *Communications Research 1948–1949* (pp. 111–129). New York: Harper & Brothers.

Berkowitz, D. (1990). Refining the gatekeeping metaphor for local television news. *Journal of Broadcasting & Electronic Media, 34*, 55–68.

Bill Moyers Journal: Buying the war. (2007). Retrieved from http://www.pbs.org/moyers/journal/btw/biosandterms.html

Bilton, N. (2013). *Hatching Twitter*. New York: The Penguin Group.

Bissinger, B. (1998). Shattered glass. Retrieved from http://www.vanityfair.com/magazine/1998/09/bissinger199809

Bleske, G. L. (1991). Ms. Gates takes over: An updated version of a 1949 study. *Newspaper Research Journal, 12*, 88–97.

Blum, L. (2015). Snapchat's new "Discover" feature brings news to millennials. Retrieved from http://college.usatoday.com/2015/02/06/snapchats-new-discover-feature-brings-news-to-millennials/

Blumberg, S., & Luke, J. V. (2013). Wireless substitution: Early release of estimates from the National Health Interview Survey, January–June 2013. Retrieved from http://www.cdc.gov/nchs/data/nhis/earlyrelease/wireless201312.pdf

Brock, A. (2012). From the blackhand side: Twitter as a cultural conversation. *Journal of Broadcasting & Electronic Media*, 56(4), 529–549.

Bruni, F. (2014). Dear millennials, we're sorry. Retrieved from http://mobile.nytimes.com/2014/06/08/opinion/sunday/dear-millennials-were-sorry.html?referrer=&_r=1

Build your ethics code. (n.d.). Retrieved from http://ethics.journalists.org/

Burton, T., & Roberts, C. (2015). *The buzz on BuzzFeed: Can readers learn the news from lists?* Paper presented at the AEJMC, San Francisco.

Buttry, S. (2013). How to verify information from tweets: Check it out. Retrieved from https://stevebuttry.wordpress.com/2013/01/21/how-to-verify-information-from-tweets-check-it-out/

BuzzFeed's newest traffic driver: Debate about the color of a dress. (n.d.). Retrieved from http://money.cnn.com/2015/02/27/media/the-dress-buzzfeed-traffic/index.html

BuzzFeed president: "We feel strongly that traditional media have given up on young people." (2013). Retrieved from http://www.theguardian.com/media/2013/oct/09/buzzfeed-facebook-al-jazeera-social-news

Cáceres, I. B. (2011). *Toward a theory on gender and emotional management in electoral politics: A comparative study of media discourses in Chile and the United States* (Ph.D. Dissertation), University of Texas at Austin, Austin, Texas.

Cameron, B., & Darwell, B. (2015). 2015 year in review. Retrieved from http://newsroom.fb.com/news/2015/12/2015-year-in-review/

Can Americans name the five freedoms in the First Amendment? (n.d.). Retrieved from http://www.newsengagement.org/home/wp-content/uploads/2015/09/Infographic-First-Amendment-2015-News-Engagement-Day-Final.pdf

Caramanica, J. (2015). A bloom in TV's Asian-American desert: In "Fresh Off the Boat," jostling cultural legacies. Retrieved from http://www.nytimes.com/2015/02/01/arts/television/fresh-off-the-boat-is-based-on-the-eddie-huang-memoir.html?smprod=nytcore-iphone&smid=nytcore-iphone-share

Carvin, E. (n.d.). ONA ethics: Social networks. Retrieved from http://ethics.journalists.org/topics/social-networks/

Cassidy, W. P. (2006). Gatekeeping similar for online, print journalists. *Newspaper Research Journal*, *27*(2), 6–23.

Chan-Olmsted, S., Rim, H., & Zerba, A. (2012). Mobile news adoption among young adults: Examining the roles of perceptions, news consumption, and media. *Journalism & Mass Communication Quarterly*, 90(1), 126–147.

Chyi, H. I., Lee, A. M., & Holton, A. (2015). Examining the third-person percception on news consumers' intention to pay. *Electronic News*, 1–21.

Clark, M. D. (2014). *To tweet our own cause: A mixed-methods study of the online phenomenon "black twitter."* (Ph.D. dissertation), University of North Carolina at Chapel Hill.

Clarke, P. (1965). Parental socialization values and children's newspaper reading. *Journalism Quarterly*, *42*, 539–546.

The Common Sense census: Media use by tweens and teens. (2015). Retrieved from https://www.commonsensemedia.org/sites/default/files/uploads/research/census_researchreport.pdf

ComScore reports February 2015 U.S. smartphone subscriber market share. (2015). Retrieved from http://www.comscore.com/Insights/Market-Rankings/comScore-Reports-February-2015-US-Smartphone-Subscriber-Market-Share

Confidence in institutions: Trends in Americans' attitudes toward government, media, and business (n.d.). Retrieved from http://www.apnorc.org/PDFs/Confidence%20in%20Institutions/ConfidenceInstitutions%20to%20DTP%20FORMATTED.pdf

Cooke, J. (1980). Jimmy's world. Retrieved from https://www.washingtonpost.com/archive/politics/1980/09/28/jimmys-world/605f237a-7330-4a69-8433-b6da4c519120/

Corasaniti, N. (2016). Snapchat bets big on quick-fire approach to campaign coverage. Retrieved from http://www.nytimes.com/2016/02/13/us/politics/snapchat-bets-big-on-quick-fire-approach-to-campaign-coverage.html?smprod=nytcore-iphone&smid=nytcore-iphone-share&_r=0

Corasaniti, N. (2015). Texting comes of age as a political message. Retrieved from http://www.nytimes.com/2015/08/19/us/politics/presidential-campaigns-see-texting-as-a-clear-path-to-voters.html?smprod=nytcore-iphone&smid=nytcore-iphone-share

Coronel, S., Coll, S., & Kravitz, D. (2015). *Rolling Stone* and UVA: The Columbia University Graduate School of Journalism report: An anatomy of a journalistic failure. Retrieved from http://www.rollingstone.com/culture/features/a-rape-on-campus-what-went-wrong-20150405

DeGusta, M. (2012). Are smart phones spreading faster than any technology in human history? Retrieved from http://www.technologyreview.com/news/427787/are-smart-phones-spreading-faster-than-any-technology-in-human-history/

Dittmar, K. (2015). 2016 outlook: Gender bias, media, and the cause for concern in presidential politics. Retrieved from http://presidentialgenderwatch.org/2016-outlook-gender-bias-media-and-the-cause-for-concern-in-presidential-politics/

The diversifying electorate—voting rates by race and Hispanic origin in 2012 (and other recent elections). (2013). Retrieved from https://www.census.gov/prod/2013pubs/p20-568.pdf

Dougherty, C., & Steel, E. (2015). YouTube introduces YouTube red, a subscription service. Retrieved from http://www.nytimes.com/2015/10/22/technology/youtube-introduces-youtube-red-a-subscription-service.html?smprod=nytcore-iphone&smid=nytcore-iphone-share&_r=0

Facebook company info. (2015). Retrieved from http://newsroom.fb.com/company-info/

Facebook reports fourth quarter and full year 2014 results. (2015). Retrieved from http://files.shareholder.com/downloads/AMDA-NJ5DZ/4008393196x0x805563/8729BE65-53AC-4B13-9F1F-F893D52D7924/FB_News_2015_1_28_Financial_Releases.pdf

Faughnder, R., & Villarreal, Y. (2015). With an eye on glowing clout of Latinos, Univision files IPO. Retrieved from http://www.latimes.com/entertainment/envelope/cotown/la-et-ct-univision-ipo-latinos-20150703-story.html

Fry, R. (2015). Millennials surpass Gen Xers as the largest generation in U.S. labor force. Retrieved from http://www.pewresearch.org/fact-tank/2015/05/11/millennials-surpass-gen-xers-as-the-largest-generation-in-u-s-labor-force/

Funke, D. (2015). We took to Twitter to ask "Would you feel safer if guns were legal on your campus?". Retrieved from http://college.usatoday.com/2015/10/10/twitter-end-campus-violence/

Global smartphone penetration 2014. (November 11, 2014). Retrieved from https://ondevice-research.com/blog/global-smartphone-penetration-2014

Gonchar, M. (2014). 50 ways to teach with current events. Retrieved from http://learning.blogs.nytimes.com/2014/10/07/50-ways-to-teach-current-events/?_php=true&_type=blogs&_r=3

Graham, J. (2015). Who's making money in digital music? Retrieved from http://www.usatoday.com/story/tech/2015/02/15/whos-making-money-in-digital-music/23176177/

Guynn, J. (2015). Facebook's Instagram hits 400 million users. Retrieved from http://www.usatoday.com/story/tech/2015/09/22/instagram-400-million-users-facebook/72637166/

Here's Donald Trump's presidential announcement speech. (2015). Retrieved from http://time.com/3923128/donald-trump-announcement-speech/

Hornaday, A. (2015). "Spotlight" joins "All the President's Men" in the pantheon of great journalism movies. Retrieved from https://www.washingtonpost.com/goingoutguide/movies/spotlight-joins-all-the-presidents-men-in-the-pantheon-of-great-journalism-movies/2015/11/12/a4e9e7a6-86ed-11e5-be39-0034bb576eee_story.html?tid=kp_google

How groups voted in 2012. (n.d.). Retrieved from http://www.ropercenter.uconn.edu/polls/us-elections/how-groups-voted/how-groups-voted-2012/

How millennials get news: Inside the habits of America's first digital generation. (2015). Retrieved from http://www.americanpressinstitute.org/wp-content/uploads/2015/08/AP_NORC_API-Millennials_Topline_REVISED.pdf

How to repay your loans. (n.d.). Retrieved from https://studentaid.ed.gov/sa/repay-loans

Howe, N., & Strauss, W. (2000). *Millennials rising: The next great generation*. New York: Vintage Books, A Division of Random House.

Hu, E. (2014). How Yahoo's diversity numbers compare with Google's. Retrieved from http://www.npr.org/sections/alltechconsidered/2014/06/17/323040120/how-yahoos-diversity-numbers-compare-to-googles?sc=17&f=1001&utm_source=iosnewsapp&utm_medium=Email&utm_campaign=app

In swing states, elusive youth voters are politically jaded and still undecided. (2012). *PBS Newshour*.

Incollingo, J. (2015). *I'm a news junkie .... I like being informed ... uses & gratifications and mobile news users*. Paper presented at the AEJMC, San Francisco.

Instagram. (n.d.). Retrieved from https://instagram.com/about/faq/

Irby, K. (2014). L.A. Times photographer fired over altered image. Retrieved from http://www.poynter.org/how-tos/writing/9289/l-a-times-photographer-fired-over-altered-image/

Isaac, M. (2014). 50 million new reasons BuzzFeed wants to take its content far beyond lists. Retrieved from http://mobile.nytimes.com/2014/08/11/technology/a-move-to-go-beyond-lists-for-content-at-buzzfeed.html?_r=0

Isaac, M. (February 17, 2015). Valuation for Snapchat could rise to $19 billion. Retrieved from http://bits.blogs.nytimes.com/2015/02/17/snapchat-said-to-seek-funding-at-up-to-19-billion-valuation/?smprod=nytcore-iphone&smid=nytcore-iphone-share

Isaacson, W. (2014). *The innovators: How a group of hackers, geniuses, and geeks created the digital revolution*. New York: Simon & Schuster.

Katz, E., Blumler, J. G., & Gurevitch, M. (1974). Utilization of mass communication by the individual. In J. G. Blumler & E. Katz (Eds.), *The uses of mass communications: Current perspectives on gratifications research* (pp. 19–32). Beverly Hills: Sage Publications.

Kent, T. (n.d.). A customized ethics code for every organization. Retrieved from http://ethics.journalists.org/about/

Kimball, P. (1959). People without papers. *Public Opinion Quarterly*, *23*, 389–398.

Kirkpatrick, D. (2010). *The Facebook effect: The inside story of the company that is connecting the world*. New York: Simon & Schuster Paperbacks.

Kludt, T. (2015). Brian Williams' suspension is over. Retrieved from http://money.cnn.com/2015/08/11/media/brian-williams-suspension-over/

Kosoff, M. (2015). 22 Millennials reveal what they really think about Snapchat Discover. Retrieved from http://www.businessinsider.com/what-millennials-think-of-snapchat-discover-2015-4

Kovach, B., & Rosenstiel, T. (2007). *The elements of journalism: What newspeople should know and the public should expect* (2nd ed.). New York: Three Rivers Press.

Kovach, B., & Rosenstiel, T. (2014). *The elements of journalism: What newspeople should know and the public should expect* (3rd ed.). New York: Three Rivers Press.

Krug, S. (2016). Reactions now available globally. Retrieved from http://newsroom.fb.com/news/2016/02/reactions-now-available-globally/

Krums, J. (2014). I took the iconic "Miracle on the Hudson" photo. Retrieved from http://mashable.com/2014/01/15/miracle-on-the-hudson-photo/#:eyJzIjoidCIsImkiOiJfZT-V4bnY5cWc3ZWpiejk4NSJ9

Lee, A. (2013). News audiences revisited: Theorizing the link between audience motivations and news consumption. *Journal of Broadcasting & Electronic Media*, *57*(3), 300–317.

Lewin, T. (1986). Winans conviction upheld. Retrieved from http://www.nytimes.com/1986/05/28/business/winans-conviction-upheld.html

Lichterman, J. (February 23, 2015). Snapchat stories: Here's how 6 news orgs are thinking about the chat app. Retrieved from http://www.niemanlab.org/2015/02/snapchat-stories-heres-how-6-news-orgs-are-thinking-about-the-chat-app/

LinkedIn. (n.d.). Retrieved from https://www.linkedin.com/about-us

Linthicum, K. (2015). Asians to surpass Latinos as largest immigrant group in U.S., study finds. Retrieved from http://www.latimes.com/nation/la-na-immigration-asians-20150928-story.html

Lynch, R., & King, S. (2015). Emmys 2015: Nominations for "Empire," "Murder" actresses make history. Retrieved from http://www.latimes.com/entertainment/envelope/tv/la-et-st-emmy-nominations-2015-story.html#page=1

Maerz, J. (2015). The best practices for reaching a Millennial audience. Retrieved from http://www.americanpressinstitute.org/publications/reports/strategy-studies/millennials-best-practices/

Maltais, M. (2015). Even young children are tech savvy, some using mobile devices before age 1. Retrieved from http://www.latimes.com/science/sciencenow/la-sci-sn-young-children-tablets-pediatrics-20151102-story.html

Marks, M., & Martin, W. (n.d.). Generation of debt: Millennials battle persistent debt, job-less woes. Retrieved from http://res.dallasnews.com/interactives/generation-of-debt/employment/

McCarty, B. (2011). The history of the smartphone. Retrieved from http://thenextweb.com/mobile/2011/12/06/the-history-of-the-smartphone/

McCombs, M., & Poindexter, P. (1983). The duty to keep informed: News exposure and civic obligation. *Journal of Communication*, *33*(2), 88–96.

McKirdy, E. (2015). China's online users more than double entire U.S. population. Retrieved from http://www.cnn.com/2015/02/03/world/china-internet-growth-2014/index.html

The media future summit. (2015). Retrieved from http://www.mediafuturesummit.com/#overview

Milbrath, L. W., & Goel, M. L. (1977). *Political participation: How and why do people get involved in politics?* (2nd ed.). Chicago: Rand McNally College Publishing Company.

"Millennials and News." (2015). Retrieved from https://www.facebook.com/Millennials-and-News-181390031899756/

The millennials are coming. (2007). Retrieved from http://www.cbsnews.com/news/the-millennials-are-coming/

Millennials: Confident. connected. open to change. (2010). Retrieved from http://www.pewsocialtrends.org/files/2010/10/millennials-confident-connected-open-to-change.pdf

Miller, E. (2015). White House hosts tribal youth, the "heart of the American story." Retrieved from http://www.npr.org/sections/itsallpolitics/2015/07/12/422317360/white-house-hosts-tribal-youth-the-heart-of-the-american-story?sc=17&f=1001&utm_source=iosnewsapp&utm_medium=Email&utm_campaign=app

Mills, K. (1988). *A place in the news: From the women's pages to the front page*. New York: Dodd, Mead & Co.

Minorities account for nearly all U.S. population growth. (2011). Retrieved from http://www.pewresearch.org/daily-number/minorities-account-for-nearly-all-u-s-population-growth/

Mitchell, A., Gottfried, J., Kiley, J., & Matsa, K. E. (2014). Political polarization & media habits. Retrieved from http://www.journalism.org/2014/10/21/political-polarization-media-habits/

Mitchell, A., Gottfried, J., & Matsa, K. E. (2015). Millennials and political news: Social media—the local TV for the next generation? Retrieved from http://www.journalism.org/2015/06/01/millennials-political-news/

Mitchell, A., Jurkowitz, M., & Olmstead, K. (2014). Social, search & direct appendix. Retrieved from http://www.journalism.org/2014/03/13/appendix-2/

Mitchell, A., Kiley, J., Gottfried, J., & Guskin, E. (2013). The role of news on Facebook. Retrieved from http://www.journalism.org/2013/10/24/the-role-of-news-on-facebook-2/

The mobile revolution: How mobile technologies drive a trillion-dollar impact. (2015). Retrieved from https://www.bcgperspectives.com/content/articles/telecommunications_technology_business_transformation_mobile_revolution/?chapter=2

Molyneux, L. (2014). Reporters' smartphone use improves quality of work. *Newspaper Research Journal*, 35(4), 83–97.

Molyneux, L. K. (2015). *Civic engagement in a mobile landscape: Testing the roles of duration and frequency in learning from news*. (Ph.D. Dissertation), University of Texas at Austin, Austin, Texas.

Most influential gadgets and gizmos (n.d.). Retrieved from http://content.time.com/time/specials/2007/article/0,28804,1638782_1638778_1638763,00.html

Murthy, D. (2013). *Twitter: Social communication in the Twitter age*. Cambridge, UK: Polity Press.

Musulin, K. (2014). Q & A with USA Today College: "Don't be afraid to take an opportunity." Retrieved from http://ajr.org/2014/03/17/q-usa-today-college-dont-afraid-take-opportunity/

Nations, D. (n.d.). What is Fremium? Retrieved from http://ipad.about.com/od/iPad-Glossary/a/What-Is-Freemium.htm

Netburn, D. (2015). Science explains how time spent outdoors colors your view of #thedress. Retrieved from http://www.latimes.com/science/sciencenow/la-sci-sn-scientific-papers-the-dress-20150513-story.html

Newspaper circulation volume. (n.d.). Retrieved from http://www.naa.org/Trends-and-Numbers/Circulation-Volume/Newspaper-Circulation-Volume.aspx

The *New York Times* passes one million digital subscriber milestone. (August 6, 2015). Retrieved from http://investors.nytco.com/press/press-releases/press-release-details/2015/The-New-York-Times-Passes-One-Million-Digital-Subscriber-Milestone/default.aspx

Nintendo. (n.d.). Retrieved from http://www.nintendo.com/corp/history.jsp

Noer, M. (2014). Read the original Forbes takedown of Stephen Glass. Retrieved from http://www.forbes.com/sites/michaelnoer/2014/11/12/read-the-original-forbes-takedown-of-stephen-glass/#1cab8b672d74

NYT VR: How to experience a new form of storytelling from the *Times*. (2015). Retrieved from http://www.nytimes.com/2015/11/08/magazine/nyt-vr-how-to-experience-a-new-form-of-storytelling-from-the-times.html?smprod=nytcore-iphone&smid=nytcore-iphone-share&_r=1

Okrent, D. (2004). The public editor: Weapons of mass destruction? or mass distraction? Retrieved from http://www.nytimes.com/2004/05/30/weekinreview/the-public-editor-weapons-of-mass-destruction-or-mass-distraction.html

Online commenting. (n.d.). Retrieved from http://ethics.journalists.org/topics/online-commenting/

Online panels. (n.d.). Retrieved from http://www.aapor.org/Education-Resources/Election-Polling-Resources/Online-Panels.aspx

Ortiz, E. (2015). Twitter's top news hashtags of 2015 include #PrayForParis, #LoveWins, #BlackLivesMatter. Retrieved from http://www.nbcnews.com/storyline/2015-year-in-review/twitter-s-top-news-hashtags-2015-include-prayforparis-lovewins-blacklivesmatter-n481081

Papper, B. (2015). Research: Minority numbers slide, women make gains. Retrieved from http://www.rtdna.org/article/research_minority_numbers_slide_women_make_gains

Parrish, M., & Nauss, D. W. (1993). NBC admits it rigged crash, settles GM suit. Retrieved from http://articles.latimes.com/1993-02-10/news/mn-1335_1_gm-pickup

Percentage of minorities in newsrooms remains relatively steady; 63 percent of newspapers have at least one woman among top-three editors. (2015). Retrieved from http://asne.org/content.asp?pl=121&sl=415&contentid=415

Perspectives on the future of news. (2012). Retrieved from http://www.aejmc.org/home/wp-content/uploads/2012/09/Sept-2012-issue.pdf

Pew Research Center. (October 2014). Mobile technology fact sheet. Retrieved from http://www.pewinternet.org/fact-sheets/mobile-technology-fact-sheet/

Pew Research Center: Millennials & political news. (2015). Retrieved from http://www.journalism.org/files/2015/06/Millennials-and-News-FINAL-7-27-15.pdf

Pinterest. (n.d.). Retrieved from https://help.pinterest.com/en/guide/all-about-pinterest

Plouffe, D. (2009). *The audacity to win: The inside story and lessons of Barack Obama's historic victory*. New York: Viking.

Poindexter, P. (2008a). Factors contributing to the sex divide in newspapers and television news. In P. Poindexter, S. Meraz, & A. S. Weiss (Eds.), *Women, men, and news: Divided and disconnected in the news media landscape* (pp. 17–34). New York: Routledge.

Poindexter, P. (2008b). Finding women in the newsroom and in the news. In P. Poindexter, S. Meraz, & A. S. Weiss (Eds.), *Women, men, and news: Divided and disconnected in the news media landscape* (pp. 65–83). New York: Routledge.

Poindexter, P. (2008c). Trouble in the news media landscape. In P. Poindexter, S. Meraz, & A. S. Weiss (Eds.), *Women, men, and news: Divided and disconnected in the news media landscape* (pp. 3–16). New York: Routledge.

Poindexter, P. (2008d). When women ignore the news. In P. Poindexter, S. Meraz, & A. S. Weiss (Eds.), *Women, men, and news: Divided and disconnected in the news media landscape* (pp. 35–47). New York: Routledge.

Poindexter, P. M. (2009). Diversity: Content. In C. H. Sterling & D. C. Whitney (Eds.), *Encyclopedia of journalism* (pp. 434–439). Thousand Oaks, CA: Sage.

Poindexter, P. M. (2011). African-American images in the news: Understanding the past to improve future portrayals. In S. D. Ross & P. M. Lester (Eds.), *Images that injure: Pictorial stereotypes in the media* (5th ed., pp. 107–120). Santa Barbara, CA: Praeger.

Poindexter, P. M. (2012). *Millennials, news, and social media: Is news engagement a thing of the past?* New York: Peter Lang.

Poindexter, P. (2014). Is news engagement endangered? Retrieved from http://www.aejmc.org/home/wp-content/uploads/2013/12/AEJMC-News_Jan-2014.pdf

Poindexter, P. M., & Conway, M. (2005). Local, network TV news shows significant gains. In E. K. Grusin & S. H. Utt (Eds.), *Media in an American Crisis* (pp. 175–186). Lanham, MD: University Press of America.

Poindexter, P., Meraz, S., & Weiss, A. S. (Eds.). (2008). *Women, men, news: Divided and disconnected in the news media landscape*. New York: Routledge.

Poindexter, P., & Harp, D. (2008). The softer side of news. In P. Poindexter, S. Meraz, & A. S. Weiss (Eds.), *Women, men, news: Divided and disconnected in the news media landscape* (pp. 85–96). New York: Routledge.

Poindexter, P., Heider, D., & McCombs, M. (2006). Watchdog or good neighbor? The public's expectation of local news. *Harvard International Journal of Press/Politics, 11*(1), 77–88.

Poindexter, P. M., & McCombs, M. E. (2001). Revisiting the civic duty to keep informed in the new media environment. *Journalism & Mass Communication Quarterly, 78*(1), 113–126.

Politics. (n.d.). Retrieved from http://www.merriam-webster.com/dictionary/politics

President Teresa A. Sullivan statement regarding Columbia University Graduate School of Journalism report. (2015). Retrieved from http://news.virginia.edu/content/president-teresa-sullivan-statement-regarding-columbia-university-graduate-school-journalism

Public service loan forgiveness. (n.d.). Retrieved from https://studentaid.ed.gov/sa/repay-loans/forgiveness-cancellation/public-service

Quick facts. (n.d.). Retrieved from http://www.census.gov/quickfacts/table/PST045215/00

Rampell, C. (2011). More college graduates take public service jobs. Retrieved from http://www.nytimes.com/2011/03/02/business/02graduates.html?_r=0

Ramsey, D. X. (2015). The truth about black twitter: Complex, influential, and far more meaningful than the sum of its social justice-driven hashtags. Retrieved from http://www.theatlantic.com/technology/archive/2015/04/the-truth-about-black-twitter/390120/

Rausch, N. (2015). Your cheat sheet on the top 2016 presidential candidates. Retrieved from http://college.usatoday.com/2015/08/26/2016-candidate-cheat-sheet/

Reese, S. D., Oscar H., Gandy, J., & Grant, A. E. (Eds.). (2001). *Framing public life: Perspectives on media and our understanding of the social world.* Mahwah, N.J.: Lawrence Erlbaum Associates.

Report of the national advisory commission on civil disorders. (1968). New York: Bantam Books.

Rieder, R. (2015a). BuzzFeed a burgeoning journalistic force. Retrieved from http://www.usatoday.com/story/money/columnist/rieder/2015/06/15/buzzfeed-expands-as-journalism-force/71250444/

Rieder, R. (2015b). The newspaper business' travails. Retrieved from http://www.usatoday.com/story/money/columnist/rieder/2015/11/02/rieder-newspaper-business-travails/75035268/

Riffkin, R. (2015). Racism edges up again as most important U.S. problem. Retrieved from http://www.gallup.com/poll/184193/racism-edges-again-important-problem.aspx?utm_source=Politics&utm_medium=newsfeed&utm_campaign=tile

Rogers, E. M. (2003). *Diffusions of innovations* (5th ed.). New York: Free Press.

The role of news on Facebook: Common yet incidental. (2013). Retrieved from http://www.journalism.org/files/2013/10/facebook_news_10-24-2013.pdf

Sebastian, M. (2015a). *New York Times* shifting resources from "every division" to mobile: Conversation with Cliff Levy, editor of NYT Now. Retrieved from http://adage.com/article/media/york-times-shifts-resources-mobile/297664/

Sebastian, M. (2015b). Publishers stare down an 'oh sh*t' mobile moment: As readers turn to smartphones, publishers struggle to monetize the traffic. Retrieved from http://adage.com/article/digital/publishers-stare-sh-t-mobile-moment/297017/

Sebastian, M. (2015c). TheSkimm, an email newsletter, has the attention of media execs and celebs: "Killing it" among millennial women. Retrieved from http://adage.com/article/media/theskimm-attention-media-execs-celebs/298534/

Severin, W. J., James W., & Tankard, J. (2001). *Communication theories: Origins, methods, and uses in the mass media* (5th ed.). New York: Longman.

Shoemaker, P. J., & Reese, S. D. (2014). *Mediating the message in the 21st century: A media sociology perspective* (3rd ed.). New York: Routledge.

Singer, J. B. (2001). The metro wide web: Changes in newspapers gatekeeping role online. *Journalism & Mass Communication Quarterly*, 78(1), 65–80.

Snapchat. (n.d.). Retrieved from http://blog.snapchat.com/page/5

Social, search, and direct: Pathways to digital news. (2014). Retrieved from http://www.journalism.org/files/2014/03/SocialSearchandDirect_PathwaystoDigitalNews.pdf

SPJ code of ethics. (2014). Retrieved from http://www.spj.org/ethicscode.asp

Stanglin, D., & Mutzabaugh, B. (March 5, 2015). Delta plane veers off runway in landing at LaGuardia. Retrieved from http://www.usatoday.com/story/news/nation/2015/03/05/plane-laguardia-skidded-off-runway-landing/24433915/

Statement of principles. (n.d.). Retrieved from http://asne.org/content.asp?pl=24&sl=171&contentid=171

Steel, E. (2015). Netflix, Amazon and Hulu no longer find themselves upstarts in online streaming. Retrieved from http://www.nytimes.com/2015/03/25/business/media/netflix-amazon-and-hulu-no-longer-find-themselves-tvs-upstarts.html?smprod=nytcore-iphone&smid=nytcore-iphone-share&_r=0

Steel, E., & Somaiya, R. (2015). Brian Williams suspended from NBC for 6 months without pay. Retrieved from http://www.nytimes.com/2015/02/11/business/media/brian-williams-suspended-by-nbc-news-for-six-months.html

Steinberg, J. (2003). Executive editor of the *Times* and top deputy step down. Retrieved from http://www.nytimes.com/2003/06/05/national/05SHELL-PAPE.html?pagewanted=all

Steinberg, J. (2004). *USA Today* finds top writer lied. Retrieved from http://www.nytimes.com/2004/03/20/us/usa-today-finds-top-writer-lied.html?pagewanted=all

Stephens, B. (2012). To the class of 2012: Attention graduates: Tone down your egos, shape up your minds. Retrieved from http://www.wsj.com/news/articles/SB10001424052702304451104577389750993890854

Stephens, M. (2007). *A history of news* (3rd. ed.). New York: Oxford University Press.

Tankard, J. W. Jr., Hendrickson, L., Silberman, J., Bliss, J., & Ghanem, S. (1991). *Media frames: Approaches to conceptualization and measurement.* Paper presented at the Association for Education in Journalism and Mass Communication, Boston.

Tea Party movement. (n.d.). Retrieved from https://en.m.wikipedia.org/wiki/Tea_Party_Movement

Thorson, E., Shoenberger, H., Karaliova, T., Kim, E. A., & Fidler, R. (2015). News use of mobile media: A contingency model. *Mobile Media & Communication*, *3*, 160–178.

Trends in consumer mobility. (2014). Retrieved from http://newsroom.bankofamerica.com/sites/bankofamerica.newshq.businesswire.com/files/press_kit/additional/2014_BAC_Trends_in_Consumer_Mobility.pdf

Trends in public attitudes about public confidence in institutions. (2013). Retrieved from http://www.norc.org/PDFs/GSS%20Reports/Trends%20in%20Confidence%20Institutions_Final.pdf

Triple Crown Winners. (n.d.). Retrieved from http://www.belmontstakes.com/history/triplecrownwinners.aspx

Tuchman, G. (1978). *Making news: A study in the construction of reality*. New York: The Free Press.

Tweedie, S. (2015). How to use Snapchat's new "Discover" feature. Retrieved from http://www.businessinsider.com/how-to-use-snapchat-discover-feature-2015-1

Twitter reports fourth quarter and fiscal year 2014 results. (2015). Retrieved from https://investor.twitterinc.com/releasedetail.cfm?ReleaseID=894844

US Airways Flight 1549. (n.d.). Retrieved from https://en.m.wikipedia.org/wiki/US_Airways_Flight_1549

USA TODAY College Collegiate Correspondent Program. (n.d.). Retrieved from http://college.usatoday.com/correspondent/

Valdez, M. (January 15, 2015). Top 8 most expensive games on iTunes. Retrieved from http://www.insidermonkey.com/blog/top-8-most-expensive-game-on-itunes-337306/

Veselinovic, M., & Clements, B. (2015). Reading this on a computer? Then you (probably) don't live in Africa. Retrieved from http://www.cnn.com/2015/02/19/africa/africa-mobile-internet/index.html

Villarreal, Y. (2015). The billennial generation: How bilingual millennials are changing Spanish-language TV. Retrieved from http://www.latimes.com/entertainment/envelope/cotown/la-et-ct-latino-millennials-univision-20150719-story.html#page=1

Weiss, A. S. (2008). Online news: Factors influencing the divide between men and women. In P. Poindexter, S. Meraz, & A. S. Weiss (Eds.), *Women, men, and news: Divided and disconnected in the news media landscape* (pp. 117–128). New York: Routledge.

Weiss, A. S. (2013). Exploring news apps and location-based services on the smartphone. *Journalism & Mass Communication Quarterly*, 90, 435–456.

Westley, B. H., & Severin, W. J. (1964). A profile of the daily newspaper non-reader. *Journalism Quarterly*, *41*(Winter), 45–50, 156.

Westlund, O. (2015). News consumption in an age of mobile media: Patterns, people, place, and participation. *Mobile Media & Communication*, *3*(2), 151–159.

White, D. M. (1950). The "Gate Keeper": A case study in the selection of news. *Journalism Quarterly*, *27*(4), 383–390.

White, K. (2015). Millennials now outnumber Boomers, Census Bureau says. Retrieved from http://www.usatoday.com/story/news/nation/2015/06/25/millennials-now-outnumber-boomers-census-says/29294241/

Willnat, L., & Weaver, D. (2014). The American journalist in the digital age: Key findings. Retrieved from http://news.indiana.edu/releases/iu/2014/05/2013-american-journalist-key-findings.pdf

Women's history month. (2015). Retrieved from https://www.census.gov/content/dam/Census/newsroom/facts-for-features/2015/cb15ff-05_womens_history.pdf

Wright, C. R. (1975). *Mass communication: A sociological perspective* (2nd. ed.). New York: Random House.

Wright, C. R. (1986). *Mass communication: A sociological perspective* (3rd ed.). New York: Random House.

Yang, M. J. (2010). *News on the go via smartphones: Users, non-users, and the relationship between new and established news media* (Ph.D. Dissertation), University of Texas at Austin, Austin, Texas.

York, C., & Scholl, R. M. (2015). Youth antecedents to news media consumption: Parent and youth newspaper use, news discussion, and long-term news behavior. *Journalism & Mass Communication Quarterly*, 92(3), 681–699.

Young adults and the Affordable Care Act. (n.d.). Retrieved from https://www.whitehouse.gov/sites/default/files/rss_viewer/fact_sheet_young_adults_may10.pdf

INDEX

A

B

C

D

E

F

G

H

I

J

K

L

M

N

O

P

Q

R

S

T

U

V

W

Y

Z

www.ingramcontent.com/pod-product-compliance
Ingram Content Group UK Ltd.
Pitfield, Milton Keynes, MK11 3LW, UK
UKHW021852100926
8556IPUK00020B/1185